IPhone 11 2020 Edition

Learn Everything You Need to Know About iPhone including tips and tricks

PHILIP KNOLL

Copyright 2019 © Philip Knoll.

This document is geared towards providing exact and reliable information in regards to the topic and issue covered. The publication is sold with the idea that the publisher is not required to render an accounting, officially permitted, or otherwise, qualified services. From a Declaration of Principles which was accepted and approved equal by a Committee of the American Bar Association and a Committee of Publishers and Associations. In no way is it legal to reproduce, duplicate, or transmit any part of this document in either electronic means or in printed format. The recording of this strictly prohibited and the document is not allowed unless with written permission from the publisher. All rights reserved.

The information provided herein is stated to be truthful and consistent, in that any liability, in terms of inattention or otherwise, by any use or abuse of any policies, processes, or directions contained within is the solitary and utter responsibility of the recipient reader. Under no circumstances will any legal the publisher for any reparation, damages, or monetary loss due to the information herein, either directly or indirectly.

Respective authors own all copyrights not held by the publisher. The information herein is offered for informational purposes solely and is universal as so. The presentation of the information is without a contract or any type of guarantee assurance. The trademarks that are used are without any consent, and the publication of the trademark is without permission or backing by the trademark owner. All trademarks and brands within this book are for clarifying purposes only and are the owned by the owners themselves, not affiliated with this document.

Printed in the United States of America

Graw-Hill Publishing House

2 Penn Plaza,

NY 10121

New York

USA

Copyright © 2019 Philip Knoll

All rights reserved.

IPHONE 11 2020 EDITION

ISB: 9781709090899

INTENSIONALLY LEFT BLANK

DEDICATION

To Philip parents, patty jean, James Knoll and my loving wife and son Diana, Kevin who are a constant source of love, encouragement, and positive energy

TABLE OF CONTENTS

1. Overview of iPhones from 2007-1999 — 1
2. How to insert SIM card — 11
3. What we expected from Apple's September event — 25
4. 25 Tips and Tricks for new iPhone 11 — 31
5. Four methods to Recover Deleted Texts from Your iPhone — 40
6. How to Download Pictures from your PC — 46
7. The things you should consider while handling your iPhone — 50
8. Important information for handling iPhone — 57
9. How to Reset an iPhone setting — 61
10. Repairing Battery and Charging issues — 69

11	Solving Privacy Problems	82
12	Troubleshooting Web Browsing Privacy Problems	93
13	IPhone 11 at a glance	98
14	Phone calls with iPhone	108
15	An iPhone Security	121
16	Travel with your iPhone	128
17	How to make emergency calls	136
18	Best iPhone 11 Screen Protectors in 2019	140

IPHONE 11 2020 EDITION

NOTE PAGE

iPhone 11

2020 EDITION

Learn Everything you Need To Know About iPhone Including Tips and Tricks

Philip Knoll

Chapter One

Introduction

I want to thank you and congratulate you on downloading this book, "iPhone 11 2020 Edition." This book contains proven steps, strategies, tips, and tricks on How to "Learn everything you need to know about iPhone."

Thanks again for buying this book. I hope you enjoy it!

Overview of iPhones from 2007-1999

Do you know when the first iPhone came out? We have the complete history of the iPhone in this chapter. The year 2019 marks the 12-year anniversary of the first iPhone produced by Apple, so we'll celebrate by reviewing the evolution of iPhone models from 2007 to date.

IPHONE 11 2020 EDITION

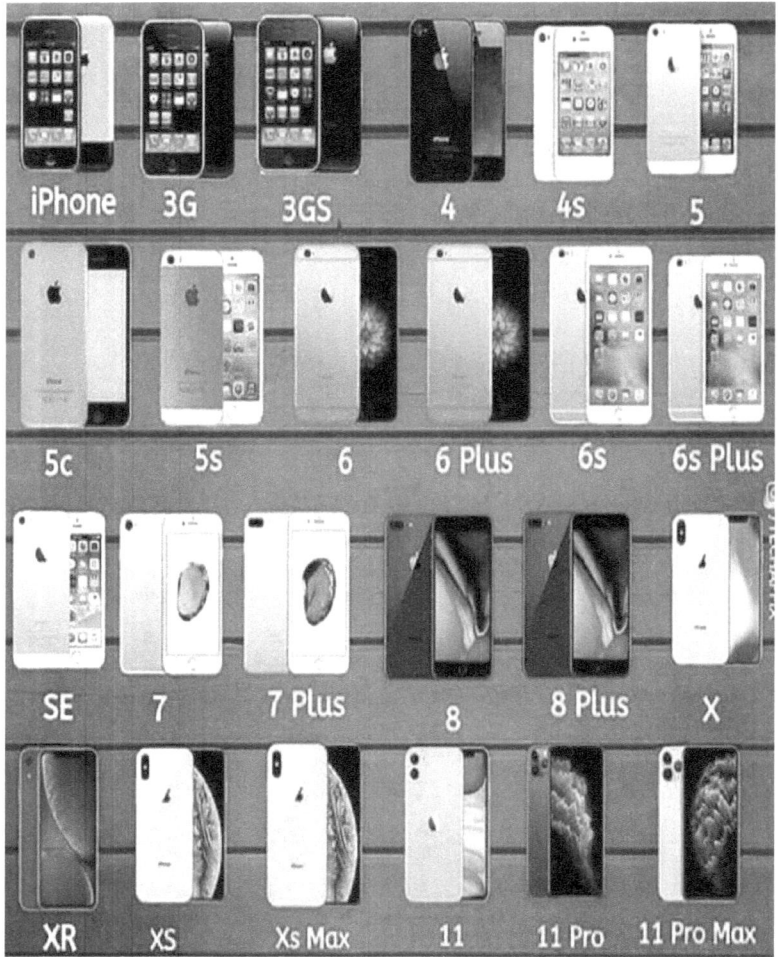

Did you know the reason why that there isn't an iPhone 2? Apple Company made the first-generation iPhone and the one that subsequently came after that was about that 3G network internet connectivity. So the iPhone 2 was skipped. Similarly, what happens to the iPhone 9? Apple capered right over that number as well and made the iPhone X. The technology giant company has produced a total of 19 iPhones from 2007 to 2019, including iPhone plus and iPhone S models, and the most recent and the latest iPhone 11, which was released in September 2019. Here is a comprehensive overview of iPhone evolution, beginning from when Steve Jobs unveiled the iPhone.

An iPhone: June 29, 2007

Can you imagine that only 16 GB is all you could put on the original iPhone? Then established there wasn't virtually as much data to put on the first iPhone yet and definitely no App Store. But you could have access to the internet on the iPhone screen you could see it on. And it could handle only 128 MB of memory. The camera was as small as 2.0 megapixels.

An iPhone 3G: July 11, 2008

The iPhone 3G wasn't big different than the original iPhone. But then there was an App Store! This iPhone got its first name for its 3G internet connectivity, which meant access to the internet you could actually browse without wanting to throw the iPhone across the room.

An iPhone 3GS: June 19, 2009

In 2009 Apple introduced the 32 GB storage capacity option with the iPhone 3GS. Evidently, introducing the App Store changed things smartly and quickly. Between music, photos, and apps, 16 GB just wasn't going to cut it all. The camera got a better upgrade to 3 MP and video recording. Apple also included Voice Control, even though we'd have to wait a couple of years before Apple introduced Siri.

An iPhone 4: June 24, 2010

The iPhone 4 release in June. Now we're getting better. The iPhone 4 was the first iPhone to come with a front-facing camera. Apple introduced a new selfies style that would take over the world. The iPhone 4 as well got a Retina display. With 512 MB memory, it was armed to handle more than even the iPhone 3GS, which only had 256 MB of memory. You can see the tech was beginning to look slightly more familiar, but 32 GB was still the maximum capacity of storage the iPhone could take.

An iPhone 4S: October 14, 2011

Talk about a gigantic difference between the iPhone 4 and the iPhone 4S: the camera went from 5 MP to a big 8 MP. Now that's an upgrade. Apple also introduced the 64 GB storage capacity option but kept the memory at 512 MB. Then Video could be recorded in 1080p. Oh, we can't forget— hello, Siri! Apple sold about four million units of the iPhone 4S in its first week of release.

An iPhone 5: September 21, 2012

This brand Apple sold more than 5 million units of the iPhone 5 in its first week of release. The camera stayed as in iPhone 4S, but the memory was improved all the way up to 1 GB. Thought 3G was cool?, the iPhone 5 had LTE connectivity. Hello, internet everywhere. Apple also introduced the new Lightning connector with the iPhone 5. And for the first time in the history of the iPhone, the screen got bigger! All previous generations' displays were made 3.5 inches, but the iPhone 5 was 4 inches.

An iPhone 5s and iPhone 5c: September 20, 2013

Between these two new devices, iPhone 5s and the iPhone 5c, Apple sold over nine million units in the first week of release. The iPhone 5c was meant to be a little more affordable and had a plastic case. It was available in four colors types, but not much else was different. The iPhone 5s, on the other hand, Apple introduced Touch ID, dual flash, and slow-motion video. Plus, it built-in the M7 motion coprocessor, which opened a new empire of possibilities and also enhanced battery life.

An iPhone 6 and 6 plus: 19, September 2014

Apple leans towards making larger leaps between the original model and the S edition than it does from the S edition to a new model. The iPhone 6's internal specs were similar to those of the iPhone 5s. The major difference was having a significantly larger screen and offering an even larger size called 6 Plus. The Retina display developed HD, and the option to get an iPhone with 128 GB of storage capacity became available. But the amount of memory was similar, and the camera didn't see a megapixel upgrade. But that didn't matter—Apple sold over 10 million units in the first week of released.

An iPhone 6s and 6s plus: September 19, 2015

Apple releases 6s and 6s Plus, and it basically considered the same. But inside, Apple upgraded the iPhone 6s relatively a lot. The camera made an enormous leap forward going from 8 MP to 12 MP. The memory was twice from 1 GB to 2 GB capacity. After the iPhone 6 had some bending issues (#bendgate), Apple furnished the iPhone 6s 7000 series aluminum to make sure that it never occurred again. This generation came with 3D Touch.

An iPhone SE: March 31, 2016

Don't you think I've forgotten with the iPhone SE. It had all the amazing internal specs of the iPhone 6s in a small package, but 3D Touch was absent. However, overall, the iPhone SE was introduced as a more affordable option that many people really valued.

An iPhone 7 and 7 Plus, September 16, 2016

Apple finally throws down the 16 GB base model option, with iPhone 7 and iPhone 7 Plus base models beginning at 32 GB of storage capacity and going up to 256 GB. Apple also introduced an innovative shiny Jet Black color design. The iPhone 7 Plus proved to be more popular than previous version Plus models due to the fact that it came with a new dual camera, which made a significantly enhanced zoom feature, and Portrait Mode, a software update that allows iPhone 7 Plus users take exciting photos using Depth of Field. Perhaps the most contentious feature of the iPhone 7 and 7 Plus was the one Apple removed: the headphone jack. The new iPhones packed with EarPods that plugged into the new Lightning port and an adaptor for regular headphones. Apple introduced its wireless AirPods, and at the same event, it announced it was removing the headphone jack.

An iPhone 8 and 8 Plus, September 22, 2017

The iPhone 8 and 8 Plus introduced with wireless charging and the glass cover on the back on the iPhone. The camera was perfect, with upgraded tools for editing and filtering images. The true display improved the viewing experience by automatically reducing blue-light exposure. Users eventually got used to not having a headphone jack and began familiarizing themselves with the wireless lifestyle.

An iPhone X, November 3, 2017

iPhone X came with awesome cameras, and the iPhone X added an extra front-facing camera that allows you to take amazing selfies in Portrait mode. Other iPhones let you take cool viewing photos, but the iPhone X included Portrait mode for the front-facing camera, and we love it at the initial aperture.

An iPhone XS and XS Max, September 21, 2018

Apple skipped the iPhone 9; Apple announced the XS and XS Max at its September 2018 event in the Steve Jobs Theater. These models were named, as they were sure to upgrade the lines of the iPhone X. Both models had the front-facing camera for Portrait-mode selfies photo. The displays were the edge to edge, and it all considered great with the Super Retina HD display. The principal upgrade might be the smallest; the A12 bionic chip increased the processing power while reducing the battery drain.

An iPhone XR, October 26, 2018

The iPhone XR was announced at the September 2018 event but wasn't available right then. Because it was inexpensive for the new models, quite a few people decided to wait for the XR to be released. This iPhone was smaller than the XS and XS Max (but bigger than 7 and 8 Plus). The display was not as crusty as the XS and XS Max, but with the Liquid Retina HD display, the difference wasn't too obvious. This model had the front-facing camera and came in more colors than the XS or XS Max.

An iPhone 11, September 20, 2019

The iPhone 11 is the newest iPhone with minimum expensive of Apple's annual series, but still has new features to be in the running for 2019's most popular iPhone produced by Apple. The device features a 6.1-inch Liquid Retina display and comes up in six crisp colors. The most thrilling addition may be the second camera on the back of the iPhone 11, both 12 MP, with ultra-wide lenses.

An iPhone 11 Pro, September 20, 2019

For some Apple customers looking for a smaller phone with a top-of-the-line display, the iPhone 11 Pro is the answer. The 5.8-inch Super Retina XDR display is Apple's most crisp and clear to date. This iPhone features not only two but also three, 12 MP HDR camera lenses, offering wide, ultra-wide, and telephoto lenses. The color options are extra muted than the iPhone 11, but there's constantly an option to add a flashy case!

An iPhone 11 Pro Max, September 20, 2019

This is the largest and most expensive iPhone of 2019, which is the iPhone 11 Pro Max. The display is Apple's Super

Retina XDR, similar to the 11 Pro, but sized at 6.5-inches. The Pro Max features similar, three-lens camera setup as the 11 Pro, as well as the same color types, making size nearly the only difference between the pro and the max devices.

An iPhone; the next Generations to Come

As you can see from the above reviewed, the iPhone has gone through several innovative changes, from a 16 GB internet web-browser to a 512 GB all-in-one camera, entertainment center, and workspace. We admired learning about the history of the iPhone! Be sure to stay up-to-date with all the newest models as the iPhone continues to evolve and grow.

What is inside the iPhone box

IPHONE 11 2020 EDITION

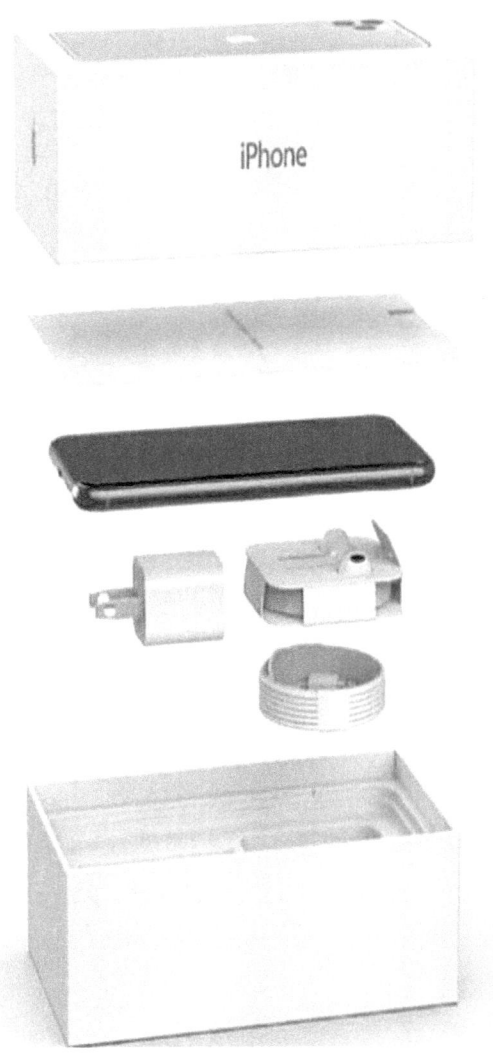

Chapter Two

An iPhone 11

How to insert SIM card

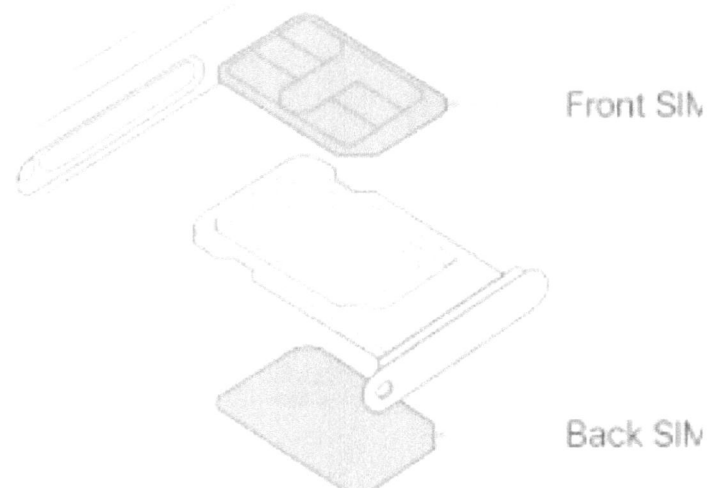

If you have set of a SIM card to install, install it before setting up the iPhone.
Tips and tricks:
A Micro-SIM card or a Nano-SIM card is a prerequisite to use mobile services when connecting to GSM networks and selected CDMA networks. iPhone that's has activated on a CDMA wireless can use a SIM card for connecting to a GSM network, predominantly for international roaming.

IPHONE 11 2020 EDITION

For iPhone 11 Pro & iPhone 11 Pro Max

Your iPhone is subject to wireless service network guidelines. It may include some limitation switching service providers and roaming, even after the conclusion of the required minimum service contract. Contact your wireless provider for more info.

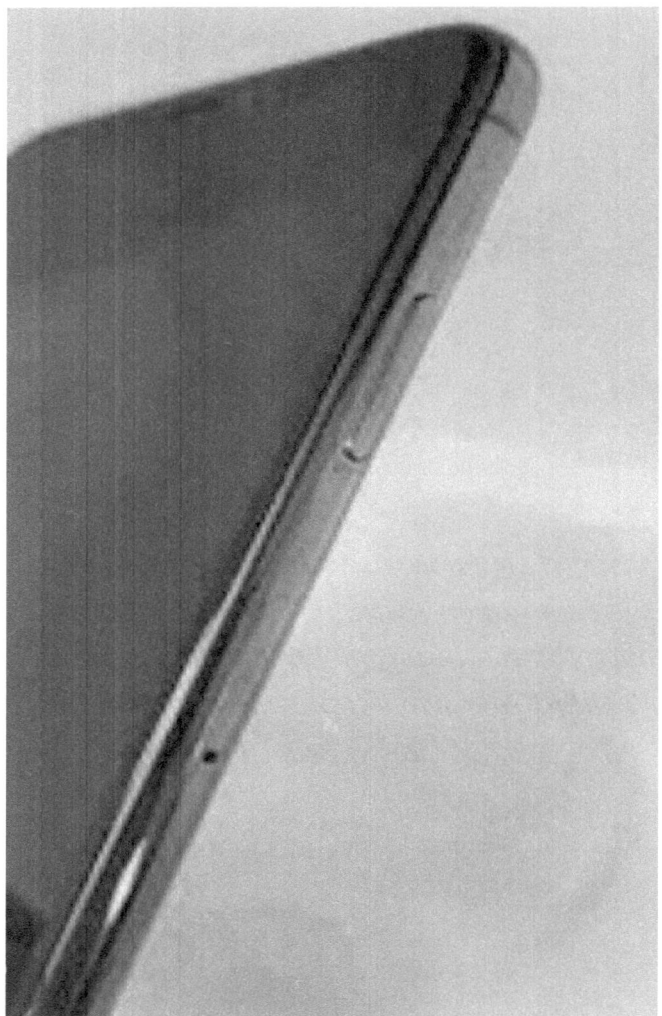

The availability of mobile capabilities depends on the wireless network.

SIM card trays, card tray Paper clip or SIM, eject tool Paper clip or SIM eject tool.

The Latest iPhone

The iPhone 11 is finally released - this is what are familiar with about it Apple has announced the iPhone 11 at its big iOS 13 launch on September 10 event in San Francisco, and we've got the first-hand information of the new phone right here.

We've checked all the key specs and features of the new iPhone as we've learned them. The iPhone 11 arrives along with the iPhone 11 Pro and iPhone 11 Pro Max, which are

IPHONE 11 2020 EDITION

Apple's top-end handsets in 2019. The iPhone 11 designed as a development to the iPhone XR.

Apple has confirmed the new iPhone 11 on stage, so we're putting all the essential information here for you to learn it.

The iPhone 11 starts at $700, which is a touch cheaper than the iPhone XR was at launch.

You can order the iPhone 11 from September 13, and you can buy it from September 20.

IPHONE 11 2020 EDITION

The new iPhone 11 design

Apple iPhone 11

SPECIFICATIONS

6.1"
828x1792 pixels

12MP
2160p

4GB RAM
Apple A13 Bionic

3110mAh
Li-Ion

The latest iPhone 11 is available in Purple, White, Yellow, Green, Black, and Product Red. The back of the iPhone is glass, and there's a Lightning connector on the bottom edge. There's no 3.5mm headphone jack on iPhone 11 that Apple really dropped the port from the new device.

It comes with a 6.0-inch Liquid Retina display, which is a similar size as the screen on the 2018 model of iPhone XR. The new iPhone 11 notch remains at the top of the screen of the device appearances remarkably like last year's phone.

The dimensions of the iPhone are 150.9mm x 75.7mm x 8.3mm, and it weighs in at 194 grams.

It's water-resistant, coming with an IP68 rating. That means you're able to submerge it in the water or in the shower, as it can survive in water up to 2 meters of water.

An iPhone 11 camera

There's a significant focus on camera this year because, s for the reason that it's comes with a dual-lens sensor, for the first time on the low-priced iPhone.

The iPhone 11 camera has a 12MP full camera with a 26mm lens, and that's pairing with a 12MP ultra front camera.

IPHONE 11 2020 EDITION

That snapper has a 120-degree field of view on the Night Mode has returned for low light surroundings, and it has come with turned on automatically on the new iPhone 11. It's similar to Google's Night Sight on Pixel phones, and the on-stage example seemed quite impressive.

Apple's example of its Night Mode

It expands and increases brightness and attempts to reduce noise in your images. Apple believes the iPhone 11 will be three times faster in low light states too.

Video modes contain 4K video recording, slow and time-lapse. Apple also says it has enhanced image stabilization on the new iPhone as well.

Overall, Apple claims that to have the "highest quality video recording in a smartphone."

The selfie front camera is a 12MP True-Depth camera, and the aim is so you can rotate the phone to take more comprehensive images even though this isn't a wide-angle shooter. It's also the first time in Apple products, and Apple has included 4K video recording on the selfie camera.

iPhone 11 specs and performance

The chipset inside the iPhone 11 is named the A13 Bionic.

Apple claims to have the fastest CPU and GPU in a smartphone.

Apple has shown off some video games played on the phone, but we've yet to study what the chipset will be capable of doing entirely.

The battery of the iPhone 11 comes with it'll last an hour longer than the iPhone XR. Apple, at no time, announces the similar sizes of its batteries, but we'll be sure to learn more it in the coming days.

The new device comes running iOS 13 software out of the box immediately that comes with a system-big dark mode for the first time.

A claim on the stage says it comes with "enhanced Face ID," but it's uncertain exactly how the facial unlocking technology has improved. The phone will be available in 64GB, 128GB, or 256GB options.

The iPhone XR was rough of a slow-burner from the previous year - it launched later than the flagship new phone, the iPhone XS, and didn't pack with a similar level of features.

Still, the fact that it was cheaper than the XS (and XS Max) means it quickly brings up traction - to the point where this year, the XR's successor, the iPhone 11, could be the crucial handset in Apple's products range.

While it doesn't have a similar feature set as the iPhone 11 Pro, the iPhone 11max require to make a splash this year - so what does it offer that the iPhone XR didn't have?

The new iPhone 11 release date is set to be earlier than its predecessor relative to the flagship phone. The iPhone XR launched later than the XS last year, but the iPhone 11 will launch alongside the iPhone 11 Pro and iPhone 11 Pro Max, with pre-orders starting on September 13 before it goes on shop sale from September 20.

Apple 2019

This is part of Apple Event, our full coverage of the latest news from Apple headquarters. September 10, Apple held its annual event, where it announced the new iPhone 11 along with other new products, including Apple Watch. The event began at 10 a.m. at the Steve Jobs Theater on Apple's campus in Cupertino, California, USA. Apple announced three new phones, replacing the iPhone XS, XS Max, and XR with the new iPhone 11, 11 Max/11 Pro, and iPhone11R. The phones likely added better cameras, fastest processors, and iOS 13. We expect to see an Apple Watch update. However, whether this is a brand new Apple Watch 5 or a minor Apple Watch upgrade remains to be seen.

Two months prior to the lunch, Apple sent invitations to the media outfit. The invitation shows that the Apple logo made up of 5 different bright colors that harken back to the old 6-color logo Apple used years back. Analysts are agitating that the unreleased iPhone 11 will not attract the buyers Apple is hoping.

Also, for the first time, Apple had stream the event live on YouTube.

The Good about the new iPhone XS, it has improved the dual camera, giving better photos than the old iPhone X in both dark and high-contrast environments. It has the fastest processor, faster face ID, adds dual-SIM support, and it's now available in gold and 512GB sub-type.

The cons of a new iPhone are its battery life is the shortest of the three new iPhones and only increases better than last year's. Despite its still-rich price tag, increased storage, USB-C fast chargers, and headphone dongles.

Considered it a latest upgrade about the X version and wait for the XR, which was released more than two months later. That was the significant feeling that holds reviewed of the XR.

The iPhone XR is Apple's most "affordable" X model and is the middle range of the new iPhone to buy right now. Starting at $750, it came with Face ID and a depth-sensing front camera instead of a usual home button, just like the rest of the iPhone X group. There's a small notch at the top of the iPhone screen. It's faster and has better battery life-span than 2017's iPhone X model, but has only a single rear camera that is equipped with its software- which assisted portrait mode that simulates depth of field effects, and a lower-res LCD screen instead of OLED. But for the vast majority of people, those are far from dealbreakers. It's the first choice you should make in buying a new iPhone.

So, where does that leave the XS and XS Max? Luxury upgrades. Everything the XS offers still stands out if you look firm enough.

An added 2x telephoto rear camera does help frame shots better and can make for more versatile portrait mode photos. The expert of the techguideblog.net has reviewed 5.8-inch iPhone XS and 6.5-inch XS Max. Both phones are an excellent device that comes with identical hardware. The 6.1-inch iPhone XR is very different enough to throw a monkey wrench into your buying plans if you can afford it.

The reason is here. The iPhone XR has one rear camera instead of two (no 2x optical zoom here) and a lower resolution LCD screen compared to the XS' OLED display. The iPhone XR also lacks the 512GB storage option of the XS models. But before you cross it off the list, note its many comparative advantages: A 6.1-inch screen that's bigger than the standard iPhone XS (and last year's iPhone X), and a battery life that Apple says will outlive the XS and XS Max.

iPhone XS

The iPhone XS in gold is a sleek, shiny iPhone
You can get a similar speedy processor in the XR as the other two iPhones, and however, there's one rear camera, which you won't allow portrait mode. That's because the XR' only single rear camera has new software to make your shots beautiful (yes, this is the attractive SLR-style "bokeh" camera effect that blurs the backgrounds of headshots). Fun, fashionable color includes a coral and blue cap of the iPhone XR's unique device personality.
 First of all, the new iPhone XR's initial price starts at

IPHONE 11 2020 EDITION

$250 less than the iPhone XS, making it the most pocket-friendly iPhone of the year 2018.

Chapter Three

What we expected from Apple's September event

The fall 2019 Apple products Special Event is only months away; on September 10, we witnessed tune in a live broadcast of the new 2019 iPhones, and more, announced from renown Steve Jobs Theater. A new set of iPhone was released as a regular yearly event that happens every year, and there's also the released for a new Apple Watch Series 5, and a new model of the Apple TV. A Bluetooth tracking device that is similar to Tile brand trackers, and the release dates for Apple's latest software updates. What we expected from Apple's September iPhone Event became true.

IPHONE 11 2020 EDITION

Apple's September 2019 Event

The first thing we noticed about the last event, the titled "By Innovation Only," is how to tune in, so we're able to view all the announcements live. As mentioned previously, the big day was September 10, 2019, and the live stream begins at 10 a.m. PT. everyone on the iPhone Life team was watching.

New 2019 iPhones released: Specs, Prices, and name

This was anticipated part of Apple's September event the new iPhone announcement, and this year we 'had expected a set of three. There have been several rumors about the specs, prices, cameras, and other features; all of these had been confirmed after the event, but let's summarize what we've observed so far.

New iPhone Names

Apple threw us for a loop when they diverged from Arabic numerals in naming the iPhone X. Will the trend continue with an iPhone XI and so on? I'd rather see a return to the familiar and a lineup called the iPhone 11, 11 Pro, and 11 Pro Max to keep things less confusing; I'm predicting that the folks at Apple will agree.

New iPhone Features

What's the big deal this year? Why does anyone need a new iPhone? For some, the decision is based on the wear and tear on their old device. Still, many are in the market for upgrades like longer battery life, faster processing, improved camera features, better internet connection, or just a more substantial, clearer display. For our feature-minded readers, here's a recap of what to expect from this year's phones.

 This year's iPhones are expected to have a 5.8, 6.1, or 6.5-inch display, and be upgrades of the XS, XR, and XS Max. This means that display size won't be changing, but don't worry, other features will be.

For those that are passionate about iPhone photography,

Apple is slated to add a second, front-facing camera to their XR upgrade. The XS and XS Max upgrades will each get a third lens, resulting in more detailed photos overall, but particularly for those where the zoom feature is used.

If you've been holding out for 5G, you may want to wait another year. According to Bloomberg, Apple and Qualcomm have had a falling out, meaning that the necessary modems for 5G service won't be available until at least 2020. A processing upgrade, involving faster speed with less battery drain, is on the horizon, though, with an A12X Bionic (used by the latest iPad Pros) or A13 chip slated for inclusion in all three iPhones.

The price of 2019 iPhone

Once you've found out whether or not you'd like to buy a new iPhone this year, the most important question to answer for

IPHONE 11 2020 EDITION

yourself is whether or not you can afford to get one. Apple's trade-in program is an excellent catch to help reduce the cost of your new iPhone device, but still... how much are we talking about the original device price? Apple used to bump up prices with every new iPhone generation, and definitely this year will be no exception. Last year (2018), the iPhone XR price started at $748, the XS at $998, and the XS Max at $1,098, an increase of about $100 from the top-of-the-line phone price of the previous year. If that trend continues, we expected to sell out around $1,200 for Apple's most recent and most significant iPhone device, and an additional hundred dollars for the other two models, as well.

Meanwhile, Apple has regularly announced a new smartwatch every September since 2015, and the romour had it that this is the year Apple watch is called Apple Watch

Series 5. As well as the 2019 iPhones, the Series 5 is keeping, 44 mm, as last year's model of Apple watches series 4. So, what new features and improvements could we expect if the size remains the same? Some new cosmetic options, like ceramic or titanium cases, will be on offer. Functional features like improved health and fitness tracking, better battery life, and a more versatile camera expected

Apple Bluetooth tracking device released

Tile Bluetooth trackers have been available for many years now, and they're beneficial for finding easy-to-lose items like purses and keys. Back on June 9-5, Mac said that Apple is looking into innovating a native Bluetooth tracker that could work with any iOS 13's to find my App.

The Apple's Bluetooth trackers will be added to your iCloud account and will display as items in the Find My App with your devices such as iPhone, AirPods, iPad, Apple Watch, and all other Apple tech. The Find My App will immensely help locate your item the tracker is paired to if it gets lost .for instance, if the tracker is put in Lost Mode, other Apple users can help you identify it, and contact you whenever it's been seen and located. Alarm notifications can be sent to your iPhone if a paired Apple Tracker is far distance, and to avoid any false alarms, then you'll be able to designate areas where it's fine to leave your tracker.

There's no detail on how much Apple Trackers will cost, but more likely they'll be more expensive than Tile products because... Apple. The Tile Pro presently retails for $35, so we'd expect the Apple equivalent to going for at least $50.

New Apple TV

On March 25, 2019, Apple announced of new services including, the Apple Card, Apple TV+, Apple News + and Apple Arcade. Apple is presently accepting applications from interested people for its credit card, and Apple News+ is open to subscribers, but what about Apple TV+ and Apple Arcade? Both of these services remain a mystery.

Apple Arcade

The notion that Apple is ready to release a new, more versatile Apple TV this year has been making the rounds. This makes a lot sense because if Apple is serious about competing in the world of gaming consoles, they need to improve the performance of their Apple TV. The logical next step would be to include the A12 Bionic or even an A13 chip in a new Apple TV.

Chapter Four

25 Tips and Tricks for new iPhone 11

Do you get an older iPhone version, learn more about the new iPhone gestures. If the new iPhone 11 is the first iPhone you use without the Home button, you'll need some time to spend in order to get used to the new iPhone gestures. But they're easy to pick up.

Start at home
Tap an app to open it.

Press the Home button anytime to return to the Home screen. Swipe left or right to see other screens.

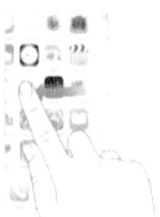

2. How to use New Direct Transfer

When you are using an iPhone 7 with iOS 12.4 and higher, you can use the new direct transfer techniques for setting up your new iPhone 11. When you want to Transfer Your Data, tap on the Transfer icon from the iPhone to transfer data wirelessly from your old iPhone to the new one. This transfer occurs on a device-to-device basis and doesn't necessarily involve iCloud. Meanwhile, the transfer will be faster, and you'll find all your apps and photos in the same state.

3. Enable the Dark Mode in your iOS 13

Surely, you're going to like the new dark mode in your iOS 13. It may flip the color scheme, giving you good a true back background with white text. Not only is it easy but also to the eyes, it helps with extending battery life. If you want to toggle dark mode, just open Control Center on the iPhone, and tap and hold on the Brightness bar. In the left corner on the bottom, you' can see a new Dark Mode toggle. Then tap on it to switch to the Dark mode instantly.

4. Learn more about Haptic Touch for iPhone 11

Apple has already removed the 3D Touch from the new iPhones. It's now replaced with Haptic Touch. That you can simply tap and hold on any object to reveal a contextual menu. In the previous version, you'd have to press a little harder on the iPhone screen.

There's a new method to rearrange apps. And previews work differently in the Safari. But, this is a welcome development. Just to tap and hold on a menu or app icon for more information.

5. How to use Ultra Wide Camera for your iPhone

Just tap on the 0.5x button to quickly switch to the ultra-wide sensor.

6. Get a Fast Charger on your device with Apple's 29W USB-C Power Adapter

The new iPhone Pro ships with an 18W charger inside the box. But unfortunately, the iPhone 11 does not. If you are using your iPhone a lot, you'll need to get a fast charger for the device. There are multiple means to do this. The easiest *ways to use the iPad's 12W power adapter if available. Because it will charge your iPhone up to twice as fast.*

If you need to do this properly, though, you have to get two things: Apple's 29W USB-C power adapter sold at ($48) and Apple's USB-C to lightning connector sold at ($28). We recommend you get out Anker 30W USB-C Wall Charger.

7. Using any Zoom Level

While the buttons give room for a precise zoom level, you can use a zoom level. The cameras will switch in the background. To do this, Just swipe on the zoom level buttons, so to reveal and dial. Then Swipe on it to change the zoom level.

8. Record Videos in 4K from the Rear Sensor

Just,Go to Settings -> Camera -> Record Video and switch to 4K.

9. Record 4K Video from Selfie Camera

From the similar settings screen, you can change the resolution for the front-facing camera to 4K as well.

10. How to take a Slofie with your iPhone

Switch to the front-facing camera and swipe over to the Slo-mo just to start taking a slow-motion video from the front-facing camera. Or as Apple called it, a slofie.

11. Zoom out ultra-wide Camera for editing

When you take a photo, the iPhone 11 captures the shot from both the wide and the ultra-wide sensor. And it keeps the ultra-wide photo around for a little time .So if like ,you can go into the editing screen and zoom out to add details that were captured from the ultra-wide photo later on.

12. Using the Night Mode features

Apple finally has a night mode in an iPhone camera, and it's actually better than what Pixel 3 has to offer previously!

While Pixel 3 takes very dramatic night mode shots, it compromises on details. The iPhone 11, in typical iPhone fashion, takes more natural-looking night time shots that are filled with details, even when it's night time. The night mode feature works automatically, and there's no button for it in the UI. It will come on automatically when the Camera app detects that you're in a low-light situation.

While Night Mode is great, it doesn't work with the ultra-wide sensor.

13. Use New QuickTime Feature

This is a new feature in the Camera app that will ship in a

few weeks after the iPhone 11 release.

You can now tap and hold on the shutter button to instantly start shooting a video, similar to how it works in Snapchat. The video will stay in the same frame and shot as the photo, which is very impressive.

If you want to keep recording the video for a longer time, you can swipe right on the shutter to lock it into video recording mode.

14. Take Burst Mode Photos

To take photos in burst mode, you now tap on the shutter button and swipe left.

15. Change Filter Intensity

When you go into the editing mode for a photo, you'll now be able to set a filtered intensity after selecting a new filter.

16. Crop and Rotate 4K Video

Tap on the edit button on a video, and you'll find new options to quickly rotate or crop a video that you took on your iPhone 11. You can also apply photo editing touches like changing the exposure and more.

17. Use New Text Editing Gestures

Apple is also taking text selection seriously. You can now just tap and hold on the cursor to pick it up and instantly move it around.

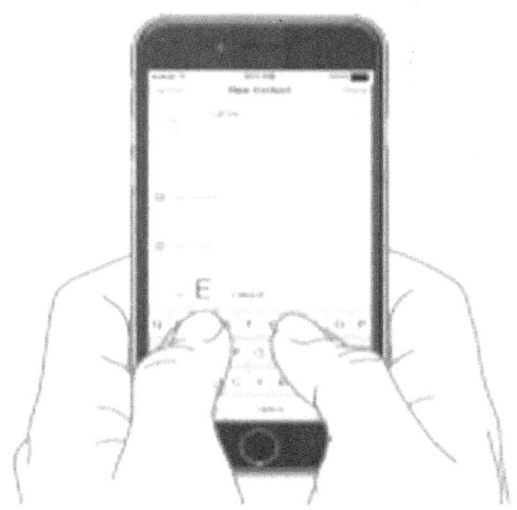

Text selection is way easier now. Just tap on a word and instantly swipe to where you want to select, like the end of the paragraph. iOS will select all the text in between the two points.

Once the text is selected, you can use gestures to copy it. Just pinch in with three fingers to copy, pinch out with three fingers to paste, and swipe back with three fingers to undo typing or action.

18. How to use Face ID from More Angles on the iPhone

Face ID in the new iPhone 11 is 30% faster. And it works more on greater angles. So even if your iPhone is not look like in front of your face, Face ID will now unlock your phone for you.

19. Gesture Typing

IPhone 11 ships with a big improvement to the keyboard. You can now swipe your finger on the letters on the keyboard to form any words. This is similar to how the Swift Key and google Gboard app works.

20. How to get The Home Button Back

iPhone X AssistiveTouch Home button

If you've migrated from the previous iPhone 7 or iPhone 8, definitely, you might be missing the Home button. But you can't get a physical Home button back if you like, there is a software called Home button on the iPhone 11 using an Accessibility setting.

To do that just, Go to Settings -> General -> Accessibility -> Assistive-touch and define shortcuts for your single tap, double-tap, long press, and 3D Touch for the Assistive-touch button. If you want to use Single Touch, define it to go Home. You can also use other gestures setting to define features you find difficult to access on the device.

21. Using Lighting Headphone Adapter for the iPhone

You might have already aware that your shiny iPhone came without a Lighting to 3.5mm headphone adapter. Apple has already bundled one on new iPhones from iPhone 7, but now, that's not the case. If you've got AirPods or you can use Apple's EarPods so far, you might not need to get the adapter.

Most people prefer to have as a backup. For instance, you might need it in your car, one when you're plugging into speakers. Apple sold lighting to a headphone adapter for about $10.

22. Shoot Photos in RAW Format Halide Featured

The iPhone 11 can takes amazing photos thanks to the new Smart HDR mode. But what if you need to take matters in your own hand? You might use an app like Halide to take images in RAW format, by using manual controls. You can set the exposure, focus, brightness, and more. The app also has an amazing Depth mode.

23. Edit Photos to Make Them Even Better

Best iPhone X Apps 7

You've got pictures from Smart HDR mode or from Halide App. They're very good, and you can make them even better using a photo editing app on the iPhone. Snapseed and Darkroom are examples. Both are easy to use. The Darkroom has a wonderful collection of filters to select from.

24. Turn off Attention Awareness

Apple iPhone uses the Face ID system to determine if you're glancing at your phone or not. And based on that, it can do several things. For instance, it can dim the screen when you're not looking at, or it can automatically silence the phone. If you don't want Apple to do these things for you, just go to Settings > Face ID & Passcode > Attention Aware Features and then turn the feature off.

25. How to Disable Tap to Wake

As we talked about above, the Tap to Wake feature can be quite useful. But it's also prone to accidental taps. If you want to disable it, go to Settings -> General -> Accessibility -> Tap to Wake.

IPHONE 11 2020 EDITION

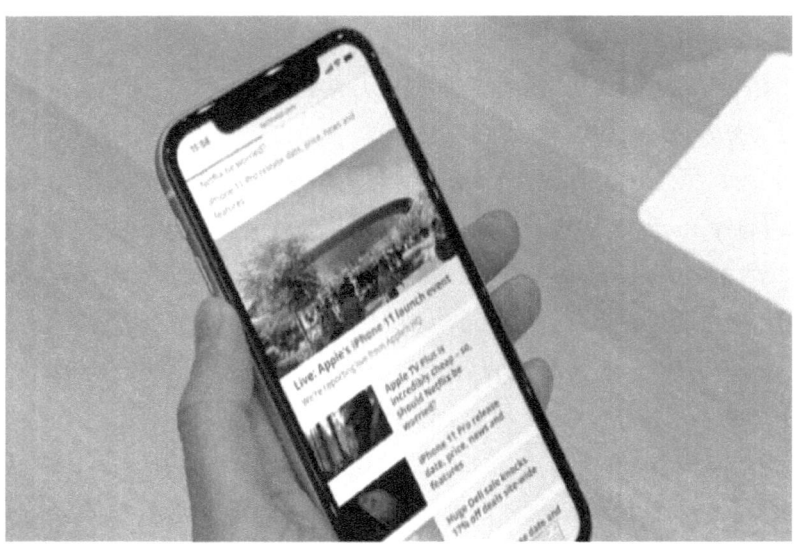

Chapter Five

Four methods to Recover Deleted Texts from

Your iPhone

When trying to get in touch with family and friends and family, unfortunately, they haven't answered your calls and messages. Are they busy? Or is there anything wrong with their device, or have they blocked your phone number? How can you know if one blocked your number? There's not a specific way to know if your number is blocked, but there are a few signs that can show someone blocked your number from texting or calling. We've discussed how to block a number from reaching you, and how to hide your number or call privately. Now let's mention the tips and tricks that can help you to calls and messages from your iPhone.

iMessage Not Delivered

If somebody has blocked you on their Phone, you won't get an alert when this happens. But, you can use iMessage to text them, but they'll never receive your message or notification of any text received in their Messages. There is one secret, you should know that you might've blocked. Look under the last text you sent before you suspect you were blocked, does it say Delivered?

If the last iMessage indicated Delivered, but the most recent one doesn't, it means that you've been blocked.

If the imessage not delivered (iMessage) Not Delivered: Use SMS to know If your Number is blocked.

If you want to see that your number has blocked, enable SMS texts on your iPhone. This means that when iMessages don't go through, your Phone re-send the message by using your data.

Imessage doesn't indicate delivered

If your SMS messages don't receive a reply or a delivery notification confirmation, it's another sign that you've blocked.

How to Know If Your Number Is Blocked: Could It Be Do Not Disturb Mode is on?

If the person that texted hasn't blocked your number and their iPhone is in Do Not Disturb mode, .then their iPhone comes on of Do Not Disturb. They'll automatically receive both a notification and iMessage.

So if your text wasn't answered, and there's no delivery notification either. There is a possibility that the Phone is out of service. Unless if there's an emergency, wait to call your friend. If your text isn't answered, after a few hours and you decide to call, here's how you can be sure you've been blocked:

I tested it, asked my brother to block my phone number, and what I discovered was that. The Phone rang, but briefly, not even a full ring tone before connecting to voicemail. I leave a message for him as I would do with an unblocked call but is display in a separate section of the voicemail on his iPhone; in the blocked calls section. After he unblocked me, my brother notified in the usual way about my voicemail on his iPhone's Lock Screen. Immediately after I was blocked, however, there was no way at all for him to know that I'd called, and no notification about the blocked voicemail. The only way he knew I rang was when he opened his voicemails and checked the Blocked Messages section.

If your number isn't blocked and the person you're attempting to call their Phone was in Do Not Disturb mode, your voicemail goes to the regular, unblocked section. Also, if you call twice within three minutes and you're not blocked, the second call may go through due to the Repeated Calls feature. The Repeated Calls feature isn't a guarantee your second call goes through, in any case.

Now that you have a few helpful tips to check if your texts and calls, you'll be able to see with reasonable assurance that your phone number has been blocked or not. The best way is to speak to the person directly, however. Probably, there's been a misunderstanding; but if they did intend to block you and no longer want to be in contact, keep it up to experience and move on.

Delete the message in your iPhone

If you have needed more space for the iPhone, you might notice that the Messages file is not enough, and you like to delete messages from your iPhone. The subsequent problem you might have is how to recover deleted text messages after you mistakenly delete texts that meant to save while trying to free up space! Fortunately, there are ways to recover deleted text messages so that you can breathe again. There are a few different methods to recover deleted text messages on your iPhone. We'll highlight how to retrieve deleted messages from iCloud Backup or, as well as getting other tips and tricks for backing those important iPhone text messages. In this chapter, you would learn a few quick and easy ways to recover deleted texts on your iPhone.

How to Retrieve Deleted Texts message and Restore with iCloud Recovery

iCloud backups are great features to recover deleted text messages from your iPhone. Have you ever backed up your Phone to iCloud storage? If yes, you should be able to use these ways to get back your deleted text messages. But, If you're not already at the moment, you should be able to do regular iCloud backups. It's a perfect idea for saving ALL of your essential information, not just messages that might potentially get deleted.

Now your iPhone has been completely erased and starts as if it's a new iPhone.

Follow this step by step instructions until you come to the Apps and Data page.

On the Apps & Data page, select Restore from iCloud Backup.

Enter your passwords for iCloud Backup, and then sign in to iCloud.

Proceed to "Choose backup," then select from a list of available backups in iCloud. There may be even some older backups to select from besides the one you have verified before starting. Select the most recent that predates the deletion of the text messages. You should keep this that the older the Backup, the more likely it does not contain essential data that you've added to your iPhone since when Backup made.

Here you must have to sign back into all your Apple accounts after the iPhone restored.

Recovery of Deleted Text Messages on iPhone and Restore from iTunes Backup

If the above method of recovering deleted text messages didn't work for your iPhone, don't be disturbed; there is another way to your aid, called iTunes. The same as recovering from a backup on iCloud, you should access any saved messages through iTunes, as long as you are doing regular backups to your PC. That is the reason why it is excellent to periodically back up your iPhone to iTunes even if you have automatic iCloud backups enabled in your device. Following these steps to recover deleted text message.

It is possible to disable this feature of the PC or Mac syncing process, so if you don't get access to your text here, that could be the reason. You may want to update these settings, mean that going forward, it is straightforward to recover deleted text messages from your iPhone.

The most important thing you to do is connect your iPhone with the PC you can sync with it. iTunes opens automatically if it's not already running on the Phone. If you want to do that, click the program and open it up. When your iPhone shows icon in iTunes, click Summary.

How to restore iPhone from iTunes

To restore the data to your iPhone comprises your deleted text messages, you need to click on "Restore Backup." If this option is to get out, you may require to change your backup selection from iCloud to your computer. You can also change this back after you have restored your iPhone.

Restore iPhone from backup recover deleted texts

This process can take a little time if you have huge data stored on your phone or Mac. If you have done a sync with your PC or Mac since deleting your text messages, this process not work that way. Because the iTunes saves only your most recent backup data for a regular restore.

For instance, if you have not done a sync with iTunes since you deleted your text messages, this option should restore those messages to your Messages app automatically!

Recover Deleted Texts by Contacting Your Service Provider

In some cases, you can recover deleted text messages by contacting your cellular network service provider. They often have access to back up your data if it has not already been overwritten or updated recently. If you need the messages back. I would recommend you to contact them before trying to use any third-party app or another drastic measure about it.

Restore Deleted Text Messages using Third-Party Apps.

Sometimes, this is the last resort for restoring your deleted text messages on your iPhone, and it's not the first choice you can use. Numerous third-party apps will help you to restore deleted documents and text messages from your iPhone.

Chapter Six

How to Download Pictures from your PC

If you want to download a photo to your pc, you can view the photos in the menu or select the Album icon on the left sidebar to view specific albums.

We're going to explain the two different methods to transfer photos with iCloud Photo Stream and iCloud Photo Library. There is some important difference between the two r; my recommendation is to enable the two.

How to Import Picture from your iPhone to Mac using iCloud Photo Stream

Photo Stream does not take any of your iCloud Storage.

However, photos will display in Photo Stream for only one month. Photo Stream doesn't upload your videos, but you can view your pictures across your devices. For both methods, Photos are automatically transferring when your device is in a Wi-Fi unless if you have enabled the option to transfer on the cellular network. To set up Photo Stream, you need to enable it on both your iPhone (or iPad) and PC.

Open the Settings app on your iPhone and then tap the banner with your name at the top of the Settings page.

How to download photos from iphone to pc
On your pc, open Preferences from the Apple menu at the top of your screen. If you're using a PC, you have to download iCloud for Windows.

How to Transfer Photos with iCloud Photo Stream
Choose iCloud and sign in.
Tap Options next to Photos.

How to transfer Picture from your iphone to computer select and Check the box next to My Photo Stream.
Then Click Done.

How to Transfer Photos with iCloud Photo Stream
Pictures automatically can be added to the Photo Stream album when you're within the Wi-Fi zone. But remember that, photos remain in Photo Stream for one month or up to

1000 images. You can set up the Photo Stream images to automatically download to your PC or Mac.

Just open the Photos app on your iPhone.

Select your Preferences at the top, below a Photo.

Then select general.

Select the box next to Importing and copy Items to the Photo Library.

This way, once photos leave your Photo Stream, they will be automatically saved to your pc. You can as well enable iCloud Photo Library on iPhone and computer if you purchase iCloud Storage. This will allow you to store photos in the Cloud, making the photos accessible online and as well as on your computer or mac.

Downloading Photos from iPhone with Airdrop on your Mac

AirDrop uses Wi-Fi connection to transfer files from iPhone to Mac, and vice versa, you should ensure that both your iPhone or iPad and pc are connected to Wi-Fi. You should also remember that Airdrop requires a 2013 Mac model or later that running OS X Yosemite to transfer files in-between Mac and iPhone.

To send files from iPhone to your Mac using AirDrop:
do the following;

Open the Photos app on the iPhone.

Select the photos you like to transfer to your pc. Tap the Share icon.

AirDrop appears at the top of the Share menu site.

Select the computer's name.

If successful, it will ask you 'Sent" your computer's name.

If your computer is not using a similar iCloud account as the iPhone you are sending files from, you are to click Save when the file arrives on your pc. If you are using similar iCloud account, the photos will automatically save

Chapter Seven

The things you should consider while handling your iPhone

Your new iPhone is made of up metal, glass, and plastic. It has sensitive electronic components. Your iPhone can damage if dropped, burned, punctured, or crushed, or if it comes in contact with liquid, chemical. Don't ever use a damaged iPhone, like one with a cracked screen, because it may cause danger or injury. If you experience the iPhone with a scratching surface, consider using a good cover case.

They are a lot of precaution you should take about your new iPhone

Repairing iPhone

Do not open the iPhone, and you should not attempt to repair or service the iPhone by yourself. Disassembling iPhone may cause an accident that may lead to injury. If your iPhone is damaged, malfunctions, or comes in contact with fluid, contact Apple or an Apple Authorized Service Provider.

iPhone battery

Do not attempt to replace your iPhone battery —you might damage the battery, which could lead to overheating or injury. The lithium-ion battery in iPhone 11 should be replaced only by an Apple Authorized Service Provider and must be disposed of separately from household waste. Do not incinerate the iPhone battery.

Any activities that may lead to distraction using an iPhone may cause a dangerous situation. Observe all rules that prohibit the use of phones or headphones (for instance, avoid using it, such as texting while driving a car or using headphones.

Charging the iPhone

Always charge yo iPhone with the included USB cable and power adapter, or with any "Made for iPhone" cables and electrical power adapters that are compatible with USB 2.0, or power adapters compliant with the following standards: EN 301489-34, IEC 62684, YD/T 1591-2009, CNS 15285, or any mobile phone power adapter with similar interoperability.

An iPhone Micro USB Adapter or other adapters may be

needed to connect your iPhone the compatible electrical power adapters. It is very important to avoid using damaged cables or charging when water moisture is present because it can result in electric shock.

When you use the compatible Apple USB Power Adapter to charge your iPhone, ensure that the AC plug / AC power cord is fully inserted correctly into the adapter prior to you plug it into an electric power outlet.

Prolonged heat exposure to your iPhone

iPhone and its power adapter always comply with surface temperature and standards limits. Even within the ranges these limits, prolonged contact with warm surfaces for a long time may result in discomfort and injury. It is better to avoid any situations where your skin is in direct contact with a device or its power supply adapter when it's plugged into a power source for a long time. For example, do not sleep or place a power adapter under your blanket or pillow when it's plugged into a main power source. It's very important to keep your iPhone and its power adapter in a securely well-ventilated area when using or to charge the device. Take special care if you have any medical condition that may affect your sensation ability to the heat against your body.

Hearing loss

Listening to sound at high decibel volumes may result in your hearing damage. All background noise pollution, as well as repetitive, prolonged exposure to high volume levels, can make sounds quieter than they actually are. It is also important to use only iPhone compatible receivers, earbuds, headphones, *speakerphones, and earpieces. Turn on the audio sound and check the volume level before inserting anything in your ear. Tips: To prevent any possible hearing damage or loss, do not listen to high volume levels for long periods of time.*

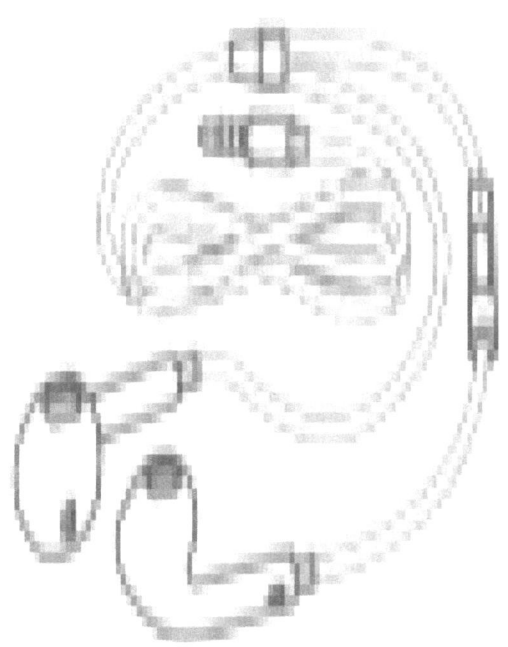

Radiofrequency interference on iPhone

For your safety, always observe any signs that indicate prohibition or restriction for the use of mobile phones (for instance, in healthcare facilities). Even though the iPhone is designed, tested, and manufactured to comply with international regulations governing radio frequency emissions, like emission from iPhone, it can affect the operation of other electronic equipment negatively, resulting in malfunction.

Use *Airplane Mode* to turn off your iPhone transmitters when using in a prohibited area, for example, while traveling in an airplane, or when asked to do so.

Medical devices and iPhone

iPhone contains a radios wave that emits electromagnetic fields and has magnets on the bottom. The headphones also have magnets in the earbuds. This electromagnetic wave and magnets may interfere with cardiac pacemakers, defibrillators, or another medical gadgets. You should at least maintain 6 inches (approximately 15 cm) distance between your pacemaker or defibrillator and your iPhone or the earbuds. If you observe your iPhone is interfering with your pacemaker or any other medical device, stop using iPhone immediately and consult your doctor for more specific information related to your medical device.

Medical conditions and iPhone

If you have any diagnosed medical condition that you could be affected by iPhone (for instance, seizures, blackouts, eyestrain, and headaches),it very imperative to consult your physician before using iPhone.

Explosive materials and iPhone

Do not charge your iPhone in any location with a potentially explosive atmosphere, such as a fueling station or in areas where the air contains chemicals or particles (like grain, dust, or metal powders). You should always obey all signs and regulations.

Repetitive motion with iPhone

When you perform repetitive activities like typing or playing games on the iPhone, you may experience some discomfort in your arms, wrists, hands, shoulders, neck, or any parts of your body. If you experience these, stop using the iPhone immediately and consult your physician.

High-consequence activities

This iPhone is not intended for use where the failure of the device may lead to death, personal injury.

Choking hazard from iPhone accessory

IPhone accessories may cause a choking hazard to small children. so it very important to be more cautions and keeps all these accessories away from small children.

Chapter Eight

Important information for handling iPhone

Cleaning an iPhone

Clean your iPhone if it comes in contact with anything that may cause stains or damage—such as dirt, ink, makeup, or lotions.

To clean iPhone:

- Disconnect all cables and turn off the iPhone (to off it just press and hold Sleep/Wake button, and slide the onscreen slider).
- You should use a soft, lint-free cloth.
- You should avoid getting moisture in openings.
- Do not use cleaning products like detergent or compressed air.

The front or back cover of your iPhone is made up of glass with a fingerprint-resistant oleophobic coating. This coating wears over a period of time with normal usage. Cleaning products and abrasive materials will increase diminish the coating and may result in glass scratch.

Using iPhone connectors, ports, and buttons

You should never force a connector into an iPhone port or apply pressure to a button, because this may result in damaging the port. If the connector iPhone and port don't join with ease, they don't match. Check for any obstructions and ensure that the connector matches the port and that you have positioned well the connector in relation to the iPhone port.

A lightning cable of the iPhone

Discoloration of the Lightning plug after use is idle. Dirt, debris, and exposure to liquids fluid may result in a color change. To remove any discoloration or if the cable is warm during use or won't charge or sync to iPhone, disconnect the Lightning cable from your PC or power adapter and clean it with a good soft, dry, lint-free cloth. Do not use liquids or cleaning agents when cleaning the Lightning connector.

Operating temperature of the iPhone

iPhone is designed and manufactured to work in ambient temp between 32° and 95° F (0°- 35°C) and stored in temperatures between -4° and 113° F (-20°- 45°C). Your iPhone can be damaged, and battery life shortened if stored or operated outside of these ambient ranges. You should avoid exposing the iPhone to unwarranted changes in temperature or humidity. When using your iPhone or charging the battery, it is normal for the iPhone to get a little warm.

If the internal ambient temperature of your iPhone exceeds normal operating temperatures (for instance, inside in direct sunlight for long periods of time), you may experience the following in attempts to regulate iPhone temperature:

- *Your iPhone stops charging.*
- *The iPhone screen dims.*
- *A temperature warning screen display.*
- *Some phone apps may close down.*

Tips and tricks:
You may not be able to use your iPhone while the temperature warning screen appears. If the iPhone can't regulate its internal ambient temperature, it goes into deep sleep mode until it cools down. Move the iPhone to a cooler environment out of direct sunlight and wait a few minutes before trying to use iPhone again.

How to restart and reset your iPhone

If something isn't working correctly, try restarting your iPhone, forcing an application to quit, or resetting the whole iPhone.

Restart iPhone.

If you want to restart your iPhone, Hold down the Sleep/Wake button until when the red slider display. Just slide your finger on the slider in order to turn off the iPhone. But if you need to turn iPhone back on, hold down the Sleep/Wake button until the Apple logo displays.

Force an application to quit.

From the home screen of your iPhone, double-click the Home button and swipe upwards on the application screen.

If you can't turn off your iPhone or if the issues continue, you may as well need to reset iPhone. But do this only if you're not able to restart the iPhone.

IPHONE 11 2020 EDITION

How to reset your iPhone.

Hold down the Sleep/Wake button tightly and the Home button at the consecutively for at least 10 seconds until the Apple logo shows up. You can also reset the dictionary, network settings, home screen layout, and location warnings sign. You can erase all of your content and settings.

Chapter Nine

How to Reset an iPhone settings

Reset iPhone settings.
If you want reset your iPhone setting, just Go to Settings > General > Reset, then select an option below:

• To Reset All Settings: All your preferences and settings are reset.

•To erase All Content and Settings: Your detail information and settings are all removed. Your iPhone cannot be used until it's set up once again.

• To Reset Network Settings: When resetting network settings, last used networks and VPN settings that weren't automatically installed, then a profile can be removed. (To delete VPN settings established by a configuration profile, just go to Settings > General > Profile, choose the profile and then tap Remove network. This can also remove other settings or associated accounts provided with the original profile.)

Wi-Fi can be turned off and then back on instantly, disconnecting the iPhone from all network you're connecting. The Wi-Fi with "Ask to Join Networks" settings always remain turned on for.

• To reset Keyboard Dictionary: You add words to the

keyboard dictionary by rejecting words iPhone suggests when you type any text. Resetting the whole keyboard dictionary erases all words you've already added.

• To reset Home Screen Layout: Returns the built-in applications to their normal layout on the iPhone Home screen.

• To reset Location & Privacy: Resets the location services and privacy settings to their factory defaults.

Getting information about your new iPhone

To see information about the iPhone. Go to Settings > General > About. The items you can view their information include:

• Network addresses
• Number of songs, videos, photos, and applications
• Capacity and available storage space
• iOS version of your iPhone
• Carrier service
• The model number of your iPhone
 • The serial number of your iPhone
• Wi-Fi and Bluetooth addresses name
• IMEI stands for' International Mobile Equipment Identity' number
• ICCID stand for (Integrated Circuit Card Identifier, or Smart Card) for GSM networks service
• MEID (Mobile Equipment Identifier) for CDMA networks
• Modem firmware
•Diagnostics and Legal (including legal notices, license, warranty, regulatory marks, and RF exposure information)

If you want to copy the serial number of your iPhone and other identifiers, touch and hold the identifier until Copy display.

To help Apple improve its products and services, the iPhone sends diagnostic and usage data instantly. This data does not specifically identify you but may include your location information. If you want View or turn off diagnostic information. Just Go to Settings > General > About > Diagnostics & Usage.

How to disabled your iPhone

If your iPhone is disabled because you forgot your passcode or entered an incorrect passcode many times, you can restore iPhone from iTunes or iCloud backup of your iPhone and reset the passcode. If you get a prompt message in iTunes that your iPhone is locked, and you must enter a passcode.

How to back up your iPhone

You can use iCloud or iTunes to back up your iPhone. If you prefer to back up using Cloud, you can't also use iTunes to back up to your PC automatically, but you can use iTunes to back up to your PC manually. iCloud backs up automatically to the iPhone daily over Wi-Fi when t's connected to a power supply and is locked. The last back up date and time is listed at the bottom of the Storage & Backup screen.

Cloud backs up the following
- Purchase music, movies, TV shows, apps, and books
- Photos and videos in your Camera Roll
- iPhone settings
- App data

- *Home screen, folders, and app layout*
- *Messages (iMessage, SMS, and MMS)*
- *Ringtone*

Tips and tricks;
Purchased content is not backed up in all-region.

To turn on your iPhone iCloud backups.

Go to Settings > iCloud, log in with your Apple ID, and password if required. Go to Storage and Backup, and then turn on iCloud backup.

To turn on backups in iPhone iTunes, just go to the File > Devices > Back Up.

Back up now.

Go to Settings > iCloud > Storage & Backup, and then tap Backup Now. Ensure you encrypt your backup. iCloud backups are encrypted automatically so that all your data is protected from unauthorized use both while it's transmitted to your iPhone and when it's stored in Cloud. If you're using iTunes for backup, choose "Encrypt an iPhone backup" in the iTunes

IPhone Summary pane.

Manage your iPhone backups.

Go to Settings > iCloud > Storage and Backup, and then tap Manage my Storage. Tap the name given to your iPhone.

You can manage which applications are backed up to iCloud, and delete existing backups. In iPhone iTunes, remove backups in iTunes Preference option.
To turn Camera Roll backup on or off.
Go to Settings > iCloud > Storage and Backup, and then tap Manage Storage. Tap the name given to your iPhone, then turn Camera Roll backup either on or off.
To View devices being backed up.
Go to Settings > iCloud > Storage and Backup > Manage Storage.
To stop iCloud backups.
Go to Settings > iCloud > Storage and Backup and then select turn off iPhone iCloud Backup.
Music not purchased in iTunes isn't backed into iCloud.
Use iTunes to back up and restore that content back to your storage.
Your last purchases may not be restored if they are no longer in the iTunes Store, App Store, or Books Store.
Tips and tricks;
Purchased and Photo Sharing content doesn't count against your GB of free cloud storage.

How to Update and restore your iPhone software

You can update your iPhone software in Settings, or by using iTunes. You can erase or restore the iPhone and then use iCloud or iTunes to restore from an iPhone backup.
Update your iPhone

You can also update iPhone software in Settings or by using iTunes.

Update wirelessly on iPhone. Just Go to Settings > General > Software Update. IPhone regularly checks for available software updates.

Update software in iTunes. iTunes regularly checks for available software updates each time you sync iPhone using iTunes.

To restore iPhone

You can use your iCloud or iTunes to restore the iPhone from a backup.

Restore from an iCloud backup.

Reset your iPhone to erase all settings and information, then sign in to iCloud and select "Restore from my iPhone Backup" in the Setup Assistant of your iPhone.

To restore from an iTunes backup.

Connect iPhone to the PC you normally sync with, choose iPhone in the iTunes window, and then click Restore in the Summary pane.

When the iPhone software is restored, you can either set it up as a new iPhone or restore your music, videos, app data, and other content from a backup.

The iPhone Cellular settings

You can use Cellular settings to turn cellular data and either roaming on or off then set which applications and services use cellular data, know the call time and cellular data usage, and set other cellular options.

If your iPhone is connected to the Internet through the cellular data network, the LTE, 4G, 3G, E, or GPRS icon display in the status bar.LTE, 4G, and 3G service on GSM cellular networks support voice and data communications consecutively. For all cellular connections, you can't use Internet services while you're talking on the iPhone unless if iPhone has a Wi-Fi connection to the Internet service. Depending on your service connection, you may not be able to receive calls while iPhone transfers data on the cellular network—when downloading a webpage for an instant.

• *GSM networks service:* On an EDGE or GPRS connection, incoming calls may directly go to voicemail during data transfers. For any incoming calls that you answer, data transfers are paused.

• *CDMA networks service:* On EV-DO connections, data transfers are paused when you answer incoming calls. On 1xRTT connections, incoming calls may directly go to voicemail during data transfers.

For any incoming calls that you may answer, data transfers are paused. Data transfer can resumes when you end your call.

If the iPhone Cellular Data is off, all data services use only Wi-Fi—such as email, web browsing, push notifications, and other services. If Cellular Data is on, charges may apply. For an instant, using certain features and services that transfer data, such as Siri and Messages, could result in charges to your data plan.

To turn Cellular Data on or off.

Go to Settings > Cellular. The following options may be available:

• *To turn Voice Roaming on or off (CDMA):* Turn Voice

Roaming off to avoid charges from using other services carriers networks. When your service carrier's network isn't available, iPhone won't have cellular (data or voice) service.
• To turn Data Roaming on or off:
Data Roaming allows Internet access on a cellular data network when you're in a location not covered by your carrier's network. When you're traveling, you can also turn off Data Roaming to avoid roaming charges.

Enable or disable 3G or 4G

Using 3G or 4G loads Internet data faster in your iPhone, but it may result in decreasing battery performance. If you're making prolonged phone calls, you may want to turn 3G or 4G off so that to extend battery life.
Set whether cellular data used for applications and other services.
Go to Settings > Cellular and then turn cellular data on or off for any application that can use mobile data. If this setting is off, your iPhone uses only Wi-Fi for that particular service. The iTunes setting contains both iTunes Match and automatic downloads from the iTunes Store or the App Store.

Sell or give away your iPhone

Make sure before you sell or give away your iPhone, try to erase all content and your personal information. If you've already enabled Find My iPhone. When activation Lock is on. You should turn off Activation Lock before the new owner can activate the iPhone under his or her account.

How to erase the iPhone and remove Activation

Lock.

If you want to remove the activation, Go to Settings > General > Reset > Erase All Content and Setting.

Chapter Ten

Repairing Battery and Charging Problems

One of the major problems of the technology world is that while our gadgets have undergone increases in performance and innovation, overall technical sophistication over the last decade or so, the battery life of those gadgets has increased comparatively slowly. Because, from the iPhone 4s to the iPhone 8 — eight generations — battery life for using the Internet over Wi-Fi rose from 10 hours to 12 hours. Apple said the iPhone 11 would get 15 hours. Still, even a 68 percent increase over eight generations is nothing proud of.

Tracking iPhone battery Use

iOS doesn't give a lot of battery data. Still, you can monitor both the total usage time (this comprises all activities: calling, surfing, playing media, etc) and standby time (a time when your iOS device is in sleep mode). One of the useful features in iOS is a breakdown of recent (the last three hours) battery usage by application, so you can see which apps have been draining your battery.

Do you want to Know Exactly How Much Battery Power You Have used

By default, iOS displays a battery icon in the status bar. As the battery drains, the amount of white inside the image gets smaller, and the level turns red when the amount of battery power gets low enough that you need to start paying attention. It is useful information, to be sure, but it's all a bit vague and imprecise.

The iOS battery icon appears in the status bar

Solution: To keep closer tabs on your device battery life, you need to tell iOS also to appear the percentage of battery power remaining. Here are the steps to follow:
 1. On the Home screen, tap Settings to open the Settings app.
 2. Tap Battery to open the Battery screen.
 3. Tap the Battery Percentage switch to On,

You Want to Know How Much You're Using Your Device

On Battery Power, Apple puts out lots of battery life numbers that include both usage mode (that is when your device is on) and standby mode (when your device is asleep). But if you need to know whether your device will have enough battery power for, say, a long plane ride or some similar extended time away from a power outlet, can you trust Apple numbers?

Tap the Battery Percentage switch On to add the percentage of battery power remaining to the status bar icon

Solution: Fortunately, you don't have to trust Apple on this because iOS keeps track of your device's overall battery use. Accurately, it tracks the amount of time since the last full charge that your device has been in usage mode and standby mode. By tracking these numbers over time and several charging cycles, you'll get to know how much battery life your iOS device gets when you use it.

Follow these steps to view these numbers:

1. On the Home screen, tap Settings to open the Settings app.
2. Tap Battery to open the Battery screen.
3. Scroll the bottom of the screen and read the Usage and Standby values.

You Suspect an App Has Been Using Too Much Battery Power

One of the benefits of adding the percentage value to the status bar's battery icon and monitoring the Usage time in the Battery screen of the Settings app is that you get to know how your apps use battery power. In particular, you might notice that your battery seems to drain a little faster than normal when you use a specific app. That's useful to know, but how can you be sure?

Solution: iOS can help by breaking down your device's battery usage by app. For both the last 24 hours and the previous 7 days, you see the percentage of total battery power that each app has used. You can also display the total amount of time each app had used the battery both when the app was onscreen, and when it was running in the background.

If you see that a particular app has been using far more battery life than you think it should – mainly if you don't use the app much more than your other apps – it might indicate a problem. For example, the app might have a memory leak, or it might be running tasks in the background.

Follow these steps to view battery usage by the app:

1. On the Home screen, tap Settings to open the Settings app.
2. Tap Battery to open the Battery screen.
3. In the Battery Usage section, tap to switch between seeing usage for the Last 24 Hours and the Last 7 Days.

. Scroll to the bottom of the Battery screen to read the total time since the last full charge that your iOS device has been in usage and standby modes

How to extend iPhone battery Life

Reducing battery consumption application as much as possible on your iOS device not only extends the time between iPhone charges but also extends the overall performance of your battery. The Battery Usage screen usually offers a suggestion or two for extending battery life, but there are many other steps you can take.

You want to prevent all Applications from Running in the iPhone Background

One of the best battery usage tools that iOS has to offer is the Time icon in the Battery screen. When activated, this feature tells you not only how much time each app has been using the battery while onscreen but, more importantly, how much time each app has been draining battery power in the background. The capability of an app to perform tasks in the background is called Background App Refresh, and it's important because although you know when an app is active onscreen, you don't always know when it's active in the background since iOS usually gives you no indication.

In practice, this means that if your battery is running low, you can stop using specific apps, but you won't know if those or other apps are still working – and therefore using up precious battery power – in the background.
Solution: You can deactivate Background App Refresh for all apps by following these steps:
 1. On the iPhone Home screen, then tap Settings to open the Settings application.
 2. Tap General to open the General screen option.
 3. Tap Background Application Refresh to open the Background App Refresh screen.
 4. Tap the Background App Refresh switch to Off, as shown in

You Want to Prevent a Specific Application from Running in the iPhone background.
 By frequent examining the battery usage of your applications – as well as an overall battery percentage and total time

onscreen and in the background – you will come to recognize any battery issues. In addition, you'll come to know which applications are consuming up your iPhone battery by running background tasks — turning off Background Application Refresh all apps.

Solution:

If you observe that a particular application is using up a higher than average percentage of battery power output in background tasks, and if you don't feel it's a basic application to run in the background, you can deactivate it, Background App Refresh for just that specific app.

Here are the steps:

1. On the iPhone Home screen, then tap Settings to open the Settings application.

2. Tap General to open the General screen.

3. Tap Background Application Refresh to open the Background Application Refresh screen.

4. In the list of applications, tap the switch to Off beside the app you no longer want to operate in the background.

. When faced with a low battery, turn off Background App Refresh to prevent any app from running in the background

You Want iPhone to Use Less Battery Power

When your iOS device battery is running low or if you still have plenty of battery power but you know you'll need to use the device for a long stretch, it would be advantageous to configure the device to use less battery power on other tasks.

Solution: You learn quite a few techniques for preserving battery power in the next section. However, iOS offers a

secure method for reducing the overall power consumption of your device. It's called Low Power Mode, and it saves battery life by doing the following:
• Turning off Background App Refresh
• Disabling the Mail app's push feature (where the app checks for new mail automatically)
• Deactivating all automatic content downloads
• Preventing all automatic app upgrade
• Disabling a few visual effects
• Dimming the screen iOS asks if you want to switch to Lower Power mode when the battery level falls to 20 percent, as.

 (This message appears again when the level drops to 10 percent.) Tap Low Power Mode to activate this feature.

You Want to Use as Low Battery Power as Possible

If battery power is scarce, there's no power outlet within reach, but you really need to use your device, then you need to minimize the amount of battery power the device uses.

Solution:

 Here are a few tips to try that should help you to reduce your device's battery consumption to a minimum, while still retaining some functionality:

• To deactivate the Background App Refresh. When you're desperate for juice, you probably don't need your applications working in the background. See the section "You want to prevent all your apps from running in the background," to learn how to turn off Background App Refresh. As an alternative, consider leaving that switch on but turning off Background App Refresh for individual apps (particularly

active apps such as Facebook and Gmail.
• Turn on Low Power Mode. Don't wait until your battery level falls to 20 percent, which is when iOS automatically offers to turn on
Low Power Mode for you. Turn it on manually earlier to increase battery life. I showed you how to activate Low Power Mode by hand in the section, "You want your iPhone to use less battery power."
•To dim the screen. The touchscreen capability drains a lot of battery power from your iPhone, so dimming it reduces the amount of power used. On the Home screen, tap Settings, tap Display & Brightness and then drag the Brightness slider to the left to dim the screen. Also, tap the Auto-Brightness switch to Off.
• To slow the auto-check on your email. Having your email regularly polls the server for new messages consumes up your battery.
To set it to check every hour or, set it to Manual check if you can. To do this, tap Settings, tap Mail, tap Accounts, and then tap Fetch New Data. In the Fetch section, tap either hourly or manually.
. Dim the screen to the lowest brightness that lets you still read the screen, and turn off the Auto-Brightness feature.
• Turn off Push. If you're not using Low Power Mode and if you have an iCloud or Exchange account, consider turning off the push feature so that to save battery power.
Tap Settings, tap Mail, tap then Accounts, and tap Fetch New Data. In the Fetch New Data, tap the Push switch to off, and in the Fetch section, then tap manually.

- *To minimize the number of applications you run.* If you won't be able to charge your device for a while, avoid background chores, such as playing music; or secondary chores, such as organizing your contacts. If your only goal is to read all your email, stick to that until it's done.
- *Make sure Auto-Lock is working.* You don't want your iOS device using up battery power while it's idle, so ensure Auto-Lock is on the job. Open Settings, tap Display & Brightness, tap Auto-Lock, and then tap a short time interval (such as 1 Minute on the iPhone or 2 Minutes on the iPad).
- *Put your iOS device into sleep mode manually, if necessary.* If you are interrupted — for instance, the pizza delivery guy shows up on time — don't wait for your device to put itself to sleep because those few seconds or minutes use battery time.

Instead, put your iPhone to sleep manually right away by pressing the Sleep/Wake button.
- *To turn off Wi-Fi if you don't need it.* When Wi-Fi is on, it frequently checks for available wireless networks, which can drain the battery. If you don't need to connect to a wireless network, turn off Wi-Fi to conserve energy. Open Settings, tap Wi-Fi and then tap the Wi-Fi
Switch to Off.
- *To turn off cellular data if you don't need it.* Your iPhone or cellular-enabled iPad always looks for nearby mobile towers to maintain the signal, which can use up battery power in a hurry. If you're surfing on a Wi-Fi network, you don't need cellular data, so turn it off. Open Settings, tap Cellular

and then tap the Cellular Data switch to Off.

   Turn off Push to ensure that this background task doesn't eat up your battery
• To turn off GPS, if you don't use it. When GPS is on, the signal receiver exchanges data with the GPS app always, which uses up iPhone battery power. If you don't need the GPS feature, for the time being, turn off the GPS antenna. Open Settings, tap Privacy, then tap Location Services, and tap the Location Services switch Off.
•To turn off Bluetooth, your iPhone, if you don't need it. When Bluetooth is running, it continually checks for nearby Bluetooth devices, and this may drain the battery so fast. If you aren't using a Bluetooth device, turn off Bluetooth so that to save energy. Open Settings, tap Bluetooth, and then tap the Bluetooth switch to Off.

Tips and tricks:
if you don't need all four of the device antennae — Wi-Fi, cellular, GPS, and Bluetooth — for a while, a faster way to turn them off is to switch your ios device to airplane mode. Either open settings and then tap the airplane mode switch to on, or swipe up from the bottom to see the Control Center and then tap the airplane mode icon.

Troubleshooting Other Battery Problems
I'll conclude this chapter with a quick look at two more

solutions to battery-related problems.

You Want to Maximize your Battery's Lifespan

It's vital to maximize battery life when you don't have nearby power, but it also essential to ensure that you maximize your battery's entire lifespan.

Solution: The lithium-ion (li-on) battery in your iOS is much more forgiving than older battery technologies such as nickel-cadmium (NiCad). With just a few simple techniques and precautions, you can ensure that your battery gives excellent performance throughout the life of your iOS device:

• Minimize the number of cycles the battery has to charge. Your iOS device battery is designed to maintain up to 80 percent of its original capacity after 500 complete discharge and charge cycles. Discharging down to zero percent and recharging counts as one cycle, discharging to 50 percent and recharging counts as half a period, and so on. Your iOS device will reach that 500-cycle

Mark slower if you plug it into a power source as often as it's convenient.

• Use the Battery. Don't think, however, that you'll keep the battery pristine by never using it. Li-ion batteries require regular use, or they'll refuse to charge.

• Don't cycle the battery. Cycling – also called re-conditioning or re-calibrating – a battery means letting it iPhone completely discharge and then fully recharging it again. It was important in earlier battery technologies, but not with the Li-on Battery in your iOS device. Letting the battery percentage drop to around 50 percent before recharging seems to be the sweet spot for maximizing battery life.

- *Avoid temperature extremes.* Exposing your iOS device to extremely hot or cold temperatures reduces the long-term effectiveness of the battery. Try to keep your iOS device at a reasonable temperature.
- *Try not to drop your iOS device.* Dropping the device can damage the battery, resulting in reduced battery performance or even eventual failure.
- *Use a high-quality charger.* Cheap chargers can damage batteries beyond repair. Either use the charger that came with your iOS device, or use a third-party charger that was designed to work with your specific make and model of the iOS device.
- *Don't store your iOS device at full charge.* If you won't be using your iOS device for a few weeks, or even longer, you'll likely be tempted to give it a full charge before putting it in storage.

However, for longer life, it's better to store the device with the battery partially discharged — to, say, between 40 and 60 percent.

Tips; it's even more important not to save your ios device with its battery fully is charged. This can damage the battery to the point where it won't charge at all when you plug the device back in.

Your battery won't Charge

When you plug your iOS device into a power outlet, you might find that it does not charge.

Solution: If you find that your battery won't load, here are some possible solutions:

- If the iPhone is plugged into a computer to charge by the

USB port, it may be that the PC has gone into standby. Waking the computer should solve the problem.
• The USB port might not be transferring enough power charges. For instance, the USB ports on most keyboards and hubs don't offer much in the way of control. If you have your iOS device plugged into a USB port on a keyboard or hub, plug it into a USB port on Mac or PC.
• Attach the USB cable to the USB power adapter and plug the adapter into an electrical power outlet.
• Double-check the connections to make sure everything plugged in properly.
• Try another Lightning cable if you have one.
• If the iOS device hasn't been charged in a long time, its battery might have gone into sleep mode to protect itself from long-term damage. In this case, it will take a few minutes — perhaps as long as ten minutes — before the battery wakes up and starts charging in the normal manner.

If none of the above suggestions solves your problem, you may need to send your device for service.

IPHONE 11 2020 EDITION

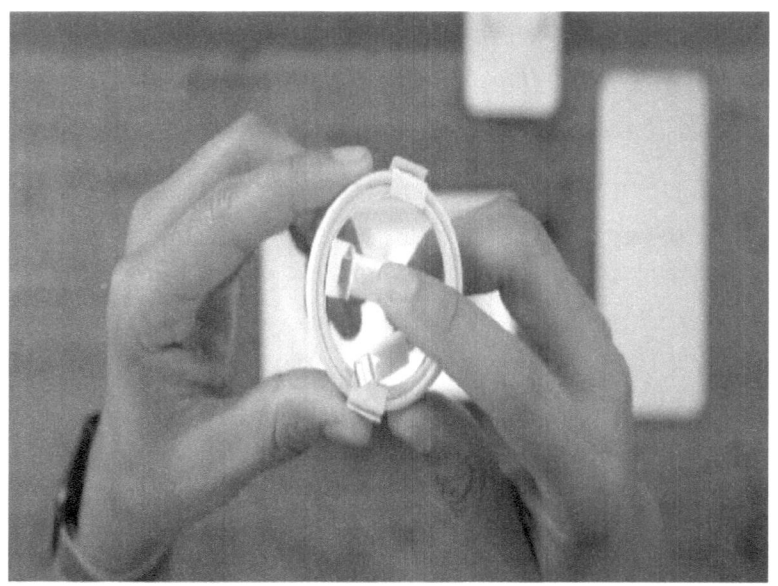

Chapter Eleven

Solving Privacy Problems

*O*ne *of the unforeseen issues of the mobile phone revolution is that we usually do our computing in* public places. Yes, it has long been people tapping on laptops in coffee shops. Still, these days, we're more likely to be touching on our phones and tablets on buses, before movies, and after classes. This means that Privacy is a potential issue because now people can easily see our screens as we work. Another unforeseen problem of the mobile device revolution is that we now carry with us a lot of personal or confidential information. Another problem related to our Privacy is because if anyone gained access to your device. That person would be at liberty to view your apps, your visited website locations, your browsing history, and much more. In the previous chapter, we learned how to lock your iPhone, but it isn't a difficult thing to experience such situation. Where someone could still get access to your device while it is unlocked. This means that it's essential to take privacy issues very seriously and to embrace a prudent paranoia: Always assume someone is watching your Screen when you're in public; assume someone could gain access to your private information.

In this chapter would learn how to take a cogent steps to solve these and similar privacy problems.

Troubleshooting General Privacy Issues

Let's begin by troubleshooting a few issues related to general privacy concerns. As You Type in your iPhone, Each Character Pops Up on the Screen, Creating a potential Privacy Risk

The iPhone onscreen keyboard comes with a feature called

character preview, which displays a pop-up version of each character as you tap it. This is great for iOS keyboard innovation because it helps them be sure they're typing accurately, but veterans often find it distracting. Either way, it's a potential privacy risk to have each character pop up when you're typing where anyone nearby can see your Screen.
Solution: Apple chose to turn off character preview by default in iOS 9, but if you see characters as you type on your iPhone, then you need to turn it off yourself by following these steps:

1. Open the Settings app.
2. Tap General.
3. Tap Keyboard.
4. Tap the Character Preview switch to Off.

Tips; even with Character preview set to on, ioS no longer shows pop-up versions of password characters. That's nice, but it still shows your most recently tapped password character for as long as three seconds! And unfortunately, there's no way to turn that off, so try to cover your password typing when you're in a public place.

You Want to Stop an Application from Using another App's Data

Third-party apps occasionally request permission to use the data from another app. For example, an app might need access to your contacts, your calendars, your photos, or your Twitter and Facebook accounts. You can always deny these requests, of course, but if you've allowed access to an app in the past, you might later change your mind and decide you'd prefer to revoke that access.

To prevent characters from popping up as you type them, turn off the Character Preview setting.

Solution: iOS offers a privacy feature that enables you to control which apps have access to your data. Here's how it works:

1. Open the Settings app.
2. Tap Privacy. The Privacy screen appears.
3. Tap the app or feature for which you want to control access. iOS displays a list of third-party apps that have requested access to the app or feature. Figure 9-2 shows an example of the Photos app.
4. To revoke a third-party app's access to the app or feature, tap its switch to off.

You Do Not Want Your Location to be tracked

In iOS, location services refer to the features and technologies that provide apps and system tools with access to the current geographical coordinates of the device. This is a simple thing, but it also needs to be kept under your control because your location data, specifically your current location, is fundamentally private and shouldn't be given out willy-nilly. Fortunately, iOS comes with a few tools for controlling and configuring a location network.

The next couple of sections would deal with how to turn off location services for individual apps as well as personal services. That fine-control is the best way to handle location services, but there may be times when you prefer a broader approach that turns off location services altogether. For example, if you're heading to a secret rendezvous (how exciting!) and you're bringing your iOS device with you, you might feel more comfortable knowing that no app or service on your device is tracking your whereabouts.

In Settings, tap Privacy and then tap an app or feature to see the third-party apps that have requested access to that item.

Tips and tricks;

On a more routine level, location services use up battery power, so if your ioS device is getting low or if you want to maximize the battery, then turning off location services will help.

Solution: Follow these steps to turn off all location services on your iOS device:

1. Open the Settings application.
2. Tap Privacy. Privacy settings appear.
3. Tap Location Services. The Location Services settings appear.
4. Tap the Location Services switch to the Off position. iOS asks you to confirm.
5. Tap Turn Off. iOS shuts off all location services

You Want to Stop an Application from using your Location.

When you open an app that comes with a GPS component, the app displays a dialog like the one shown in Figure 9-4 to ask your permission to use the GPS hardware in your device to determine your current location. Notice that iOS only allows apps to access your area while you use the app. Once you exit the app, it can no longer access your location. Tap Don't Allow if you think that your current situation is none of the app's business, or tap Allow if that's just fine with you.

To prevent your location from being used by any app or service, set the Location Services to switch to Off.

A slightly different scenario is when an app must use your location to function. A good example is Foursquare, which requires your site to show you nearby businesses and to let you "check-in" to those places. In this case, iOS automatically gives the app access to your location while you're using the app, but the app might request access to your area even when you're not using it, as shown in

Again, tap Don't Allow if you think the app is overstepping its bounds, or tap Allow if it's all good.

When you first launch a GPS-aware application, it asks your permission to access your current location while you use the app

An app that must access your area while you use the app might also seek permission to access your area when you're not using it.

Whatever type of permission you choose, after you make your decision, you might change your mind. For example, if you deny your location to an app, that app might lack some basic functionality. Similarly, if you allow an app to use your area, you might have second thoughts about compromising yo116

Solution: Whatever the reason, you can control an app's access to your location by following these steps:

1. Open the Settings app.
2. Tap Privacy. Privacy settings appear.
3. Tap Location Services. The Location Services screen

appears.

4. Tap the application in which you want to configure access to GPS. The app's location access options appear.
the options for the Foursquare app.

5. Tap one of the following options to configure the app's access to your location:

• *Never.* Tap this option if you want to deny your current location to the app.

• *While Using the App.* Tap this option if you're going to allow a specific app to access your current area only when you are actively using the application.

• *Always.* Tap this option if the app requires your location to function even when you're not using it.

(You should know that this option is only available for apps that require full-time access to GPS.)

You Want to Stop One or More System Services from using Your Location.

The iOS provides location services to various internal system services that perform different tasks, such as calibrating the compass, setting the time zone, and serving up Apple Ads that change depending on location data. You might prefer that iOS not provide your location to one or more of these services.

Use these options to configure an app's access to your location.

Solution: If you don't want iOS providing your location to some of these services, you can prevent this by following these steps:

1. Open the Settings app.
2. Tap Privacy. The Privacy screen appears.
3. Tap Location Services. The Location Services screen appears.
4. Tap System Services. iOS displays the System Services screen.

Tips and tricks:
It's also essential to know when a system service is using your location. To set this up, scroll to the bottom of the System Services screen and tap the Status bar icon switch to on. ioS now use the status bar to display one of the icons shown above this switch whenever a system service is using your location.

You want to prevent iOS from Storing a List of Your Frequent Locations

iOS keeps track of the physical locations you visit most frequently, and it offers this data to apps such as Maps and Calendar. Enables these apps to make suggestions based on your location history. Still, you might prefer that iOS not keep track of your frequent locations for privacy reasons.

Solution: You can enhance your privacy not only by clearing the list of many locations but also by preventing iOS from maintaining this list at all. Here are the steps to follow:
1. Open the Settings app.
2. Tap Privacy. The Privacy screen appears.
3. Tap Location Services. The Location Services screen appears.

4. Tap System Services. iOS displays the System Services screen.

5. Tap Frequent Locations to open the Frequent Locations screen.

6. To remove the current list of frequent locations, tap Clear History, and then when asked to confirm, tap Clear History once again.

7. To prevent iOS from storing your oft-used site, tap the Frequent Locations switch to Off.

You Do Not like to Share Your Location with your Family and Friends

When you set up Family Sharing on your iCloud account, one of the setup screens asks you about if you want to share your location with your family using the Messages and Find My Friends applications. If you initially decided to share your area, you might later decide to change your mind and keep your location private.

Solution: You can disable this feature by following these steps:

1. Open the Settings app.
2. Tap Privacy. The Privacy screen appears.
3. Tap Location Services. The Location Services screen appears.
4. Tap Share My Location to open the Share My Location screen.
5. Tap the Share My Location switch to Off.

You Do Not Want Your Device Usage

Information Being Sent to Apple.

iOS continuously monitors your device resources to watch out for adverse events. Then can include memory getting too low, processor usage getting too high, an app crashing, or the system spontaneously rebooting. When it detects such events, iOS records the current system state and writes this data to a diagnostics file.

To no longer allow family and friends to see your location, tap the Share My Location switch to off

You Don't Want to Receive Targeted Ads on your iPhone

In the same way that, online advertisers can track you across the Web using cookies (small text files that the advertisers store on your computer).Applications advertisers can follow your interests using a piece of data called the Advertising Identifier. An anonymous device identifier that iOS uses when you perform specific actions, such as searching the App Store. Advertisers have access to the Advertising Identifier. They can use it to serve you ads that are selected based on your usage. You might prefer not to receive these targeted ads.

Solution: You can configure your privacy settings to tell advertisers not to use the Advertising Identifier to track your interests and actions. You can also reset the Advertising Identifier value, which is similar to deleting the tracking cookies on your computer.

Follow these steps:
 1. *Open the Settings app.*
 2. *Tap Privacy. The Privacy screen appears.*
 3. *Tap Advertising. The Advertising screen appears.*

4. Tap the Limit Ad Tracking switch to On, .

5. Tap Reset Advertising Identifier and, when iOS asks you to confirm, tap Reset Identifier.

Another way that iOS serves you targeted ads is via your location. Here are some steps to follow to turn off this privacy breach:

1. Open the Settings app.
2. Tap Privacy. The Privacy screen appears.
3. Tap Location Services. The Location Services screen appears.
4. Tap System Services. iOS displays the System Services screen.

Figure 9-13. To prevent advertisers from using the Advertising Identifier to send you targeted ads, activate the Limit Ad Tracking setting.

5. Tap the Location-Based Apple Ads switch to Off.

You Do Not like to Be Shown Apps That Are Popular in Your Area

The App Store uses your location to tell you which apps are the most popular near you. You might prefer not to target in this way due to privacy concerns.

Solution: Follow these steps to turn off this tracking feature:

1. Open the Settings app.
2. Tap Privacy. The Privacy screen appears.
3. Tap Location Services. The Location Services screen appears.
4. Tap System Services. iOS displays the System Services screen.

5. Tap the Popular Near Me switch to Off, as shown in Figure 9-15.

Tap Location-Based Apple Ads to Off to prevent seeing targeted ads based on your location. To prevent iOS from using your location to determine popular nearby apps, tap the Popular Near Me switch to Off.

Chapter Twelve

Troubleshooting Web Browsing Privacy Problems

This chapter takes you through troubleshooting techniques related iPhone 11 to enhance the privacy of your web browsing sittings.

You like to delete the List of Websites You've Visited

Safari's History list – the group of sites you've recently surfed – is a great feature when you need it, and it's a safety feature when you don't. Though, there are times when the History list is plain straitlaced. For instance, if you visit any private corporate site, a financial website, or other location you wouldn't like others to see, the History list might betray you. And sometimes, unpleasant sites can end up in your History list by accident. For instance, you might tap a legitimate-looking link in a web page or email message, only to end up in some dark, nasty web neighborhood. , you high-tail it out of there right away with a quick tap of the Back button, the lousy site is now creeping around in your history.

Tips and trick;
Take caution as of iOS 9, clearing your Safari history also removes all your cookies and website data. Clearing cookies

might bring problems since many of them store site login data or site customizations. Hence, when clearing your history, consider removing recent data only (such as data from the previous hour browsing).

Solution: Whether you've got sites on the History list that you wouldn't need anyone to see, or you discover the idea of Safari tracking your movements on the Web to be a bit threatening, follow these steps to wipe out the History list:

1. In Safari, tap the Bookmarks button. Safari opens the Bookmarks list.
2. Tap Back till you get to the Bookmarks screen.
3. Tap History. Safari opens the History screen.
4. Tap Clear. Safari ask how much of your history you want to clear
5. Tap a time: The Last Hour, Today, Today and Yesterday, or All-Time. Safari removes every site from the History list for that time.

Tips and tricks: Safari uses your iPhone history as well as your bookmarks to analyze every page you view and determine the most likely link you'll tap — the so-called top hit — and preloads that link. If you do tap that link, the page loads lickety-split. However, if you're not at ease having Safari send your history and bookmarks to apple, you can turn this feature off. To off this feature, go to Settings, tap Safari, and then tap the preload top hit a switch to off.

You Do Not like Safari to Show Suggestions When Searching

Another technique Safari might compromise your online privacy is by displaying suggestions as you enter search text into the address bar. If someone is glancing over your shoulder or borrows your device for a quick search, she might see these suggestions.

Solution: To turn off these suggestions, follow these steps:
1. Open the Settings app.
2. Tap Safari to open the Safari screen.
3. Tap both the Search Engine Suggestions switch and the Safari
Suggestions switch it to off.

You like to browse the Web without Storing Data about the Sites You have visited

If you find yourself continuously deleting your browsing history or website data, you can save it much time by configuring Safari to do this automatically. The termed private browsing, and it means that Safari doesn't save any data as you browse.

Correctly, it doesn't save the following:
• Sites not added to the history (even though the Back and Forward buttons still work for navigating places that you've visited in the session).
• Web page text and images not saved.
• The search text not protected with the search box.
• AutoFill passwords not saved.

Solution: To activate private browsing, do the following steps:
1. In Safari, tap the Tabs button.
2. Tap Private. Safari generates a separate set of tabs for private browsing.
3. Tap Add Tab (+). Safari creates a new private tab.

Once you're done browsing privately, tap the Tabs icon, and then tap the Private Button to turn off Private Browsing.

You like to Ensure Online Promoters are not tracking you

Under the appearance of providing you with the "benefit" of targeted ads, online advertisers and promoters use cookies to track the websites you visit, the searches you conduct. This data isn't linked to you personally, but no one likes to follow in this manner. Luckily enough, iOS Safari is configured by default to curtail this kind of tracking. Still, you might want to confirm – or even strengthen – these settings.

Solution: Preventing online tracking comprises two things. First, make sure Safari's Do Not Track setting is activated, which tells sellers not to track you online. Note, however, that it does not force promoters not to follow you. Compliance is voluntary, but you should activate the setting in any case for the few advertisers who do honor it.

The second way of preventing online tracking, you need to decide the level at which you like to block cookies. You have four choices:

• Always Block. This level states Safari not to accept any cookies. But, I don't recommend this level because it disables functionality on several websites (for instance, a site's ability to save your login data and customizations).

• Allow from Present Website Only. This level makes Safari only to accept cookies that are set by whatsoever website you are currently visiting. No other website – in particular, no online advertising site – can set a cookie. If you like to

strengthen Safari's ad-blocking, this is the setting to use, though there's a chance you might lose some functionality.

• *Allow from Websites I Visit.* This level conveys Safari to only accept cookies not only from the current website but from any website you've visited in the past. For instance, suppose you have earlier visited the Facebook site. If the current location wants to set a Facebook cookie, then this setting allows it. If you've never attended a place in the past – which the case for the most majority of online advertising sites – then Safari blocks cookies from that site. The default setting, it's a good because sites you've earlier visited might need to access a cookie to implement some functionality on your current website, such as account data.

• *Always Allow.* This level makes Safari accept every cookie from any site. Avoid this setting because it means that any online advocate who does not honor the Do Not Track setting, which is. Unfortunately, the majority of them use cookies to track you.

Follow these steps to configure Safari to certify online advertisers are not tracking you:

1. Open the Settings app.
2. Tap Safari to open the Safari phone screen.
3. If needed, tap the Do Not Track switch to On.
4. Tap Block Cookies to open the Block Cookies screen.
5. Tap the cookie-blocking setting you wish to use.

Chapter thirteen

IPhone 11 at a glance

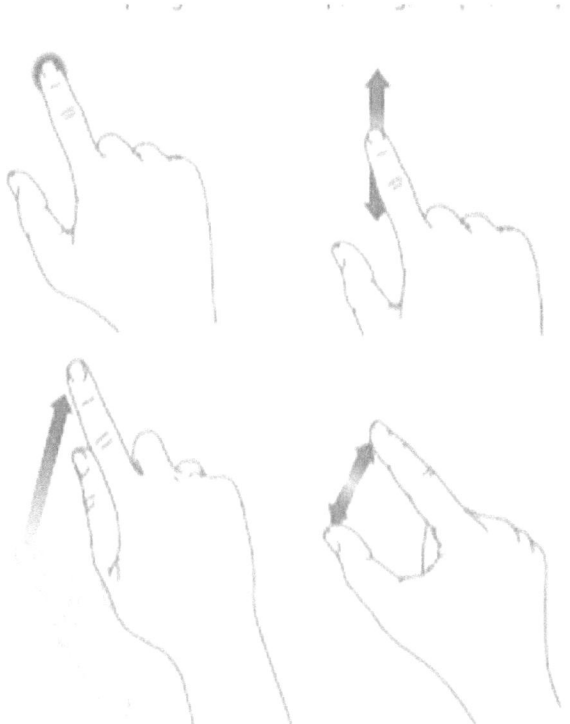

Multi-touch screen

Buttons

Most of the buttons you always use with the iPhone are virtual ones on the touchscreen. A few physical buttons control essential functions only, such as turning the iPhone on or adjusting the volume, etc.

Sleep/Wake button

When you're not using your iPhone, press the Sleep/Wake button to lock iPhone. Locking the iPhone sets the display to sleep, saves the battery, and checks anything from happening if you touch the screen. But, you can still get phone calls, Face Time calls, text messages, alarms, and notifications. You can listen to music and adjust the volume.

Sleep/Wake button sleep/Wake button

On past iPhone models, the Sleep/Wake button is on the top edge:

Sleep/Wake button sleep/Wake button iPhone locks if you don't touch the screen for a few minutes. If you want to adjust the timing, Go to Settings > General > Auto-Lock.

Turn on your iPhone. Press and hold the Sleep/Wake button up to the time Apple logo displays.

To unlock your iPhone. Press the Sleep/Wake or Home button, and then drag the slider.

Turn iPhone off. You should Press and hold the Sleep/Wake button until the slider show, then drag the slider. For more security, you can require a passcode to unlock your iPhone. Go to Settings > Touch ID & Passcode (iPhone with Touch ID) or Settings > Passcode .

The home button of your iPhone

The Home button navigates you to the Home screen and offers other convenient shortcuts. On the Home screen, tap any app to open it.

To see apps you've opened. Double-click the Home button when iPhone is unlocked.

On iPhone models with Touch ID, you can also use the sensor in the Home button to read your fingerprint as an alternative of using your passcode or Apple ID password to unlock iPhone or make any purchases in the iTunes Store, App Store, and iBooks

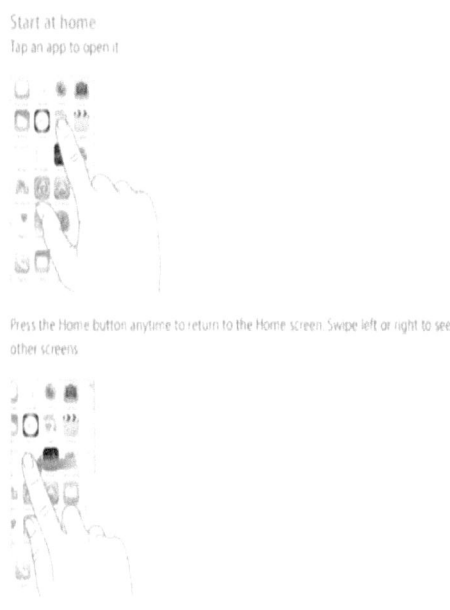

Store. If you have iPhone 6 or iPhone 6 Plus, you can use the Touch ID sensor for verification when using Apple Pay to make a purchase in any store or from within an app.

Volume controls

When you're on the call or listening to songs, movies, or other media, you can use the buttons on the side of the iPhone 11 to adjust the audio volume. Else, the buttons control the amount for the ringer, alerts, and other sound effects.

Volume up and Volume down

Lock the ring tone and alert volumes. Go to Settings > Sounds, then turn off Change with Buttons. To reduce and limit the volume for music and videos, go to Settings > Music > Volume Limit.

Tips and tricks:

In European Union (EU) countries, iPhone may warn you that you're setting the volume above the EU recommended decibel level for your hearing safety. To increase the capacity above this level, you may need to release the volume control slightly. To limit the maximum headset volume to an average level, go to Settings > Music > Volume Limit, then turn on EU Volume Limit. To avoid changes to the volume limit, go to Settings > General > Restrictions.

I am using Control Center to adjust the volume. When iPhone is locked or when you're using a different app, swipe up from the bottom edge of the screen to open the Control Center.

You can use either a volume button to take a picture or record a video with your iPhone.

Ring/Silent switch the Ring/Silent switch to put the iPhone in ring mode or silent mode.

In the ring mode, the iPhone plays all sounds. But in silent mode, the iPhone doesn't ring or play alerts and other sound effects (on the other hand, the iPhone may still vibrate).

Tips and tricks:
Clock alarms, audio apps such as Music, and several games play sounds through the built-in speaker, even when the iPhone is in silent mode. In some r, the sound effects for Camera and Voice Memos performed, yet if the Ring/Silent mode switch set to silent.

Use Do Not Disturb.

You can likewise silence calls, alerts, and notifications using Do Not Disturb. Swipe up from the bottom edge of the screen to open Control Center, then tap.

Status icons

The iPhone icons in the status bar at the top of the screen give a lot of information about iPhone:

Status icon	
E	EDGE
GPRS	GPRS/1xRTT
Wi-Fi	Wi-Fi call
🛜	Wi-Fi
☾	Do Not Disturb
⊚	Personal Hotspot
↻	Syncing
✳	Network activity

Status icon		
•••••		Cell signal
✈		Airplane mode
LTE		LTE
4G		UMTS
3G		UMTS/EV-DO

Status icon what it means phone signal when you're in range of the cellular network, and you can make and receive calls.

IPHONE 11 2020 EDITION

Status icon	
E	EDGE
GPRS	GPRS/1xRTT
Wi-Fi	Wi-Fi call
📶	Wi-Fi
🌙	Do Not Disturb
◉	Personal Hotspot
↻	Syncing
✳	Network activity
↪	Call Forwarding
[VPN]	VPN
☎	TTY
🔒	Portrait orientation lock
⏰	Alarm
➤	Location Services
✱	Bluetooth®
🔋	Bluetooth battery
▬	Battery

If there's no signal, "No service" appears. Airplane mode is on—you can't make any phone calls, and other LTE your network's LTE network is available. iPhone can connect to the Internet over that network. But this service is not available in all regions.)

UMTS your carrier's 4G UMTS (GSM) or LTE network is possible, and iPhone can also connect to the Internet over that network. (Not available in all regions.)

Status icon what it means EDGE your carrier's EDGE (GSM) network is available, and the iPhone can also connect to the Internet over that network.

GPRS/1xRTT Your carrier's GPRS (GSM) or 1xRTT (CDMA) network is available, and iPhone may connect to the Internet over that network.

Hotspot

Turn sharing action options on or off. Tap and tap More (drag options to the left if necessary). Turn of third-party sharing or action options.

Organize sharing and action options. Tap, and then tap More (drag options to the left if necessary). Touch and drag to rearrange your options.

Continuity of the iPhone

The Continuity features connect iPhone with iPad, iPod touch, and Mac so they can work together as one. You can also start an email or document on the iPhone, for instance, then pick up where you left off on your iPad. Or allow your iPad or Mac to use the iPhone to make phone calls or send SMS or MMS text messages.

Continuity features need iOS 8 or OS X Yosemite, and work with iPhone 6 or later, iPod touch (6th generation) or later, iPad (5th generation) or later, and supported Mac PCs.

Hand of Pick up on one device where you left of on another. You can also use Hand of with Mail, Safari, Pages, Numbers, Keynote, Maps, Messages, Reminders, Calendar, Contacts, and even third-party apps. For Hand of to work, your devices must be signed in to iCloud using the similar Apple ID, and it must be within Bluetooth range of one another (about 10 meters).

Switch devise.

Swipe up from the screen bottom-left edge of the Lock your screen (where you see the applications activity icon), or go to the multitasking screen, and tap the application. On Mac, open the app you were using on the iOS device.

Disable Handoff on our devices. Go to Settings > General > Hand of & Suggested Apps. Disable Handoff on our Mac. Go to System Preferences > General, and then turn off Let Hand of among this Mac and your devices set up with iCloud.

Chapter Fourteen

Phone calls with iPhone

When you want to make and receive phone calls on your iPad, iPod touch, or Mac with iOS 8 or OS as long as your iPhone is on the same Wi-Fi network, and signed in to iCloud and Face Time with the similar Apple ID. (If available on your iPhone, Let W i-Fi Calls must be of. Go to Settings > Phone > Wi-Fi Calls.)

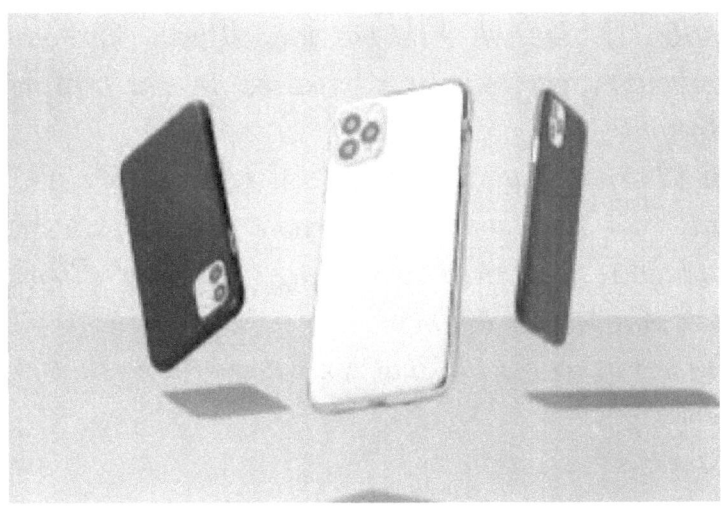

Make a phone call on your Mac, iPad, iPod touch. Tap or click a phone number in the Contacts, Calendar, or Safari. On your iPad or iPod touch, you can also tap a recent contact on the multitasking screen.

Disable iPhone Cellular Calls. Go to Settings > Face Time, then turn off iPhone Cellular Calls.

Messages Switch between your iOS devices and Mac PCs (with iOS 8 or OS X Yosemite) when you send and receive SMS and MMS text messages. Sign in to iMessage with a similar Apple ID as your iPhone.

Instant Hotspot

You can use Instant Hotspot on iPhone to provide Internet access to your other iOS devices and Mac PCs (with iOS 8 or OS X Yosemite) that are signed in to iCloud using similar Apple ID. Instant Hotspot uses iPhone Personal Hotspot, without having to enter a password or even turn on Personal Hotspot.

Use Instant Hotspot. Go to Settings > Wi-Fi on your iOS DeviceDevice, and then simply select your iPhone network under Personal Hotspots. On your Mac, select your iPhone network from your Wi-Fi settings.

When you're not using the Hotspot, your devices disconnect to save battery life.

Tips and tricks:

This feature may not be available in all-region.

Customize iPhone

Arrange your applications
Arrange applications. Touch and hold any application on the Home screen till it jiggles, and drag all apps around. Drag any app to the edge of the iPhone screen to move it to a Home screen, or to the Dock at the far bottom of your screen. Press the Home button firmly to save your new arrangement.
Create a new Home screen. Whereas arranging apps, drag an app to the right edge of the last Home screen. The dots above the Dock display how many Home screens you get, and which one you're currently viewing.

You can also customize the Home screen with iTunes when the iPhone is directly connected to the PC. In iTunes, select the iPhone and then click Apps.
Start over. Go to Settings > General > Reset and then tap Reset Home Screen Layout so that to return the previous Home screen and applications to their layout. All folders are removed, and the original wallpaper is reinstated.

Organize folders

Create a folder. When arranging apps, drag one app against another. Tap the name of the folder if you want to rename it. Drag apps to add or delete them. Press the Home button

when finished, you can have several pages of apps in one folder.
Delete a folder. Just Drag out all the apps—the folder is deleted automatically.

Change the wallpaper

Wallpaper settings allow you to set a photo or an image as wallpaper for the Lock screen or Home screen. You can select from either dynamic or still images if you like.
Change the wallpaper. Go to Settings > Wallpaper >Select a New Wallpaper.
When selecting an image for new wallpaper, the Perspective Zoom button controls your selected wallpaper is zoomed. For wallpaper, you already set, go to the Wallpaper setting, and tap the image of the Lock screen to see the Perspective Zoom button.

Tips and trick:
The Perspective Zoom button doesn't display if Reduce Motion is turned on.

Adjust the screen brightness

Dim your screen to extend battery life, or use Auto-Brightness. Adjust the screen brightness. Go to Settings > Display & Brightness, and then drag the slider. If your Auto-Brightness is on, iPhone adjusts the screen brightness for current light conditions using the built-in ambient light sensor. You can adjust the brightness in Control Center.
Display Zoom with iPhone 6. You can also expand the screen display. Go to Settings > Display & Brightness. Tap View (below Display Zoom), choice Zoomed, and then tap

Set.

Type text onscreen keyboard allows you to enter text when required.

Enter text Tap a text field on the onscreen keyboard, and tap letters to type. If you ever touch the wrong key, you can also slide your finger correct key. The letter isn't entered till you release your finger from the key.

Tap Shift to type a text like uppercase, or touch the Shift key and slide to the letter.

Double-tap Shift for caps lock. To enter numbers, punctuation, or symbols, just tap the Number key or the Symbol key. If you haven't added kind of keyboards, tap to switch to the emoji keyboard. If you have various keyboards, then tap to switch to the last one you used. Continue tapping to access any enabled keyboards, or touch and hold, and then slide to select another new keyboard.

To quickly end a sentence with a full stop and space, double-tap the space bar.

To type any alternate character, touch and hold a key, and then slide to select one of the options.

If any word show underlined in red, tap it to see suggested corrections. If the word you want doesn't display, type the correction.

As you write, the keyboard can predict your next word. Tap a word to select it, or accept a highlighted prediction by entering a space or punctuation. When you tap a recommended word, space displays after the word. If you enter a comma, period, or some punctuation, space is automatically removed. Reject suggestion by tapping your original word that shown as the predictive text option with quotation marks).

Predictive text

To hide predictive text. Pull down the suggested words. Drag the bar up when you like to see the suggestions again.

Turn off predictive text. Touch and hold or, and then slide to Predictive. If you turn off predictive text, your iPhone may still try to suggest corrections for misspelled words.

You should accept correction by entering a space or punctuation, or by tapping return. To reject a correction, tap the "x." If you throw away similar suggestions a few times, iPhone stops suggesting it.

To set options for typing or add keyboards. Go to Settings > General > Keyboard. The onscreen keyboard for iPhone 6 and later comprises additional keys you may find useful. You can also see these keys when you hold an iPhone in landscape orientation. You can as well use an Apple Wireless Keyboard to enter text.

Edit text

Revise the text. Touch and hold the text display the magnifying glass and drag to position the insertion point of the text.

Select text. Tap the insertion point to show the selection options. Or double-tap a word to select it. Drag the grab points to choose more or less text. In read-only documents, like webpages, touch and hold to choose a word.

Grab points

You can copy, cut, or paste over selected text. With some

apps, you also can also get bold, italic, or underlined text (tap B/I/U); get the definition; or have iPhone recommend an alternative. Tap to see all the options.

Undo the last edit. Shake iPhone, and then tap Undo.
Save keystrokes a shortcut allows you to enter a word or phrase by typing just a few characters. For instance, type "omw" to enter "On my way!" That one's already set up, but you can add your own.
Create a shortcut. Go to Settings > General > Keyboard and then tap Shortcuts. Have a word or phrase you can use and don't want it corrected? Create a shortcut, but let the Shortcut to field a blank

Use iCloud to keep iPhone personal dictionary up to date on your other devices. Go to Settings > iCloud, and then turn on iCloud Drive or Documents & Data.

Use an Apple Wireless Keyboard

You can use an Apple Wireless Keyboard, separately to enter text on your iPhone. The keyboard connects through Bluetooth, so you must have first pair it with Phone.

Tips and tricks:
 The Apple Wireless Keyboard may not support keyboard features that are on your Device. For instance, it does not anticipate your next word or automatically correct misspelled words.

Pair an Apple Wireless Keyboard with your iPhone.
To turn on the keyboard, go to Settings > Bluetooth and turn on Bluetooth, and then tap the keyboard when it displays in the devices list.
Once it's paired, the keyboard reconnects to your iPhone when it's in the range—up to about 10 meters. When it's connected, the onscreen keyboard doesn't display.
Save your batteries. Turn off Bluetooth and the wireless keyboard when not in use. You can also turn off Bluetooth in Control Center. To turn off the keyboard, hold down the On/Off switch until the light goes off.
Unpair to a wireless keyboard. Go to Settings > Bluetooth, then tap next to the keyboard name, and tap forget this DeviceDevice.
Add or change keyboards styles
You can turn typing features style, such as spell checking, on or off; add

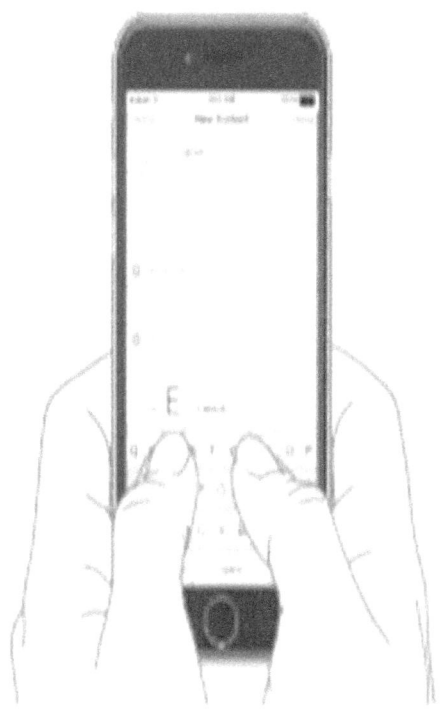

Keyboards for writing in other languages; and modify the layout of your onscreen keyboard or Apple Wireless Keyboard.

Set typing features. Go to Settings > General > Keyboard.

Add a keyboard for another language. Go to Settings > General > Keyboard > Keyboards > and select Add New Keyboard.

Switch keyboards. If you haven't added keyboards, tap to switch to the emoji keyboard.

If you get some keyboards, tap to switch to the last one you used.

If you want to continue tapping to access other enabled

keyboards, or touch and hold, and then slide to select a different keyboard.

Change the keyboard layout. Go to Settings > General > Keyboard > Keyboards, choice a keyboard, and then select a layout.

Dictate If you like, you can also dictate instead of typing. Ensure Enable Dictation is turned on the iPhone (in Settings > General > Keyboard), and the iPhone is connected to the Internet.

Keychain on an incoming call: Press and hold the center button for about three seconds, and then let go. You would get two slow beeps, which confirm declined your call.

• *Switch to an incoming or on-hold call and put the current call on hold:* Press the center button. Then press again to switch back to the first call

• *Switch to an incoming or on-hold call, and end the current call:* Press and hold the center button for about three seconds, then let go. Two low beeps confirm that you completed the first call.

Bluetooth devices

You can also use Bluetooth devices with your iPhone, as well as headsets, car kits, stereo headphones, or an Apple Wireless Keyboard.

Tips and tricks: The use of some accessories with your iPhone may affect wireless performance. Not all iPod and iPad accessories are entirely compatible with the iPhone. Turning on Airplane Mode may reduce audio interference in-between the iPhone and accessory. Reorienting or relocating iPhone and the connected accessory may increase wireless performance.

Turn Bluetooth on or off. Go to *Settings > Bluetooth.* You can as well turn Bluetooth on or off in Control Center.

Connect to a Bluetooth device. Tap the device in the iPhone devices list, then follow the on-screen guidelines to connect to it. See the detail that came with the device for information about Bluetooth pairing.

iPhone must be within about 34 feet (10 meters) of the Bluetooth device.

How to return audio output to your iPhone. Turn off or unpair the device, turn of Bluetooth in *Settings > Bluetooth,* or use AirPlay to switch the audio output to iPhone.

Ways to bypass Bluetooth. To use the iPhone receiver or speaker for any phone calls from iPhone:

- *Answer a call by tapping anywhere on the iPhone screen.*
- *During a call, tap Audio and select iPhone or Speaker.*
- *Turn off the Bluetooth, unpair it, or travel out of network range.*
- *Turn off Bluetooth in iPhone Settings > Bluetooth. If you want to unpair your device. Go to Settings > Bluetooth, then tap next to the device, and then tap Forget this Device. If you don't see the devices list, ensure that the Bluetooth is on.*

Restrictions of third –party usage

If you wish, you can set limits for some apps, and purchased content. For instance, parents can restrict explicit music from appearing in playlists, or prohibit changes to some settings. Use restrictions to limit the use of certain apps, the installation of new apps, or changes to accounts or the volume limit.

To turn on restrictions. Just Go to Settings > General >

Restrictions, then tap Enable Restrictions. You'll be asked to define restrictions passcode that's required to change the settings you make. This can be different from the passcode for unlocking your iPhone.
Tips and trick:
If you dare forget your restrictions passcode, you must restore the iPhone software.

Privacy setting

Privacy settings allow you to control which apps and system services have access to Location Services, and to contacts, calendars, reminders, and photos.

Location Services allows location-based apps such as Reminders, Maps, and Camera to collect and use data indicating your location. Your location is determined using available information from service network data, local Wi-Fi networks (if you iPhone Wi-Fi turned on), and GPS (may not be available in all regions). The area data collected by Apple isn't received in a form that personally recognizes you. When an app is using Location Services, it appears in the status bar.

Turn Location Services on or off. Now Go to Settings > Privacy > Location Services. You can also turn it off for some or for all apps and services. If you turn off Location Services, you're just prompted to turn it on again the next time on application or service tries to use it.

Turn Location Services off or system services. Several system services, such as compass calibration and location-based ads, use location or area. To see their status, turn them on or off, or show in the status bar when these services use your location, go to Settings > Privacy > Location Services > System Services.

Turn off access to private info. Go to Settings > Privacy. You can also see which apps and features have requested and been granted access to the following information:
- *Your contacts*
- *Calendar*
- *Reminders*
- *Photos*
- *Bluetooth Sharing*
- *Microphone*
- *Camera*
- *HomeKit*
- *Health*
- *Motion Activity*
- *Twitter*
- *Facebook*

You can turn off each app's access to each category of information. Review the terms and conditions for each third-party app to understand how it uses the data it's applying for.

Chapter Fifteen

An iPhone Security

The security features of your iPhone support you to protect the information from being accessed by others. Use a strong passcode with data protection for better security; you can set a passcode that must be correctly entered each time you turn on or wake up your iPhone.

How to set a passcode.

Go to Settings > Touch Apple ID & Passcode (iPhone

models with Touch ID) or Settings > Passcode and then set a four-digit passcode.

Setting a passcode turns on data protection, so using your passcode as a key to encrypt Mail messages and attachments stored on your iPhone, using 256-bit AES encryption. (Other apps may use data protection.)

Enhance security. Turn off Simpler Passcode and use a secure passcode. To enter a strong passcode that's a combination of Roman numbers and letters, you use the keyboard. If you select to unlock iPhone using the numeric keypad, set up a longer passcode using numbers only.

Add your fingerprints and set options for the Touch ID sensor. (This apply only to iPhone models with Touch ID) Go to Settings > Touch ID & Passcode.

How to access features when iPhone is locked. Go to Settings > Touch Apple ID & Passcode (iPhone models with Touch ID) or Settings > Passcode (other models). Optional features include:

- Today
- Notifications
- Siri
- Passbook
- Reply with Message

Go to Settings>Touch your ID & Passcode (iPhone models with Touch ID only) or Settings > Passcode (other models), and then tap Erase Data. After TEN failed passcode attempts, all iPhone settings are reset, and your information and media are deleted by removing the encryption key to the data.

If you dare forget your passcode, you must restore the iPhone software.

The touch ID

On any iPhone models with Touch ID, you can also unlock the iPhone by placing a finger on the button. Touch ID also allows you to do the following:

•you can use Apple ID password to make purchases in the iTunes Store, App Store, or iBooks Store.

• to resent your credit card or debit card when making any purchase in a store that deals with Apple Pay as a method of payment

• Provide your debit and credit card information, billing and shipping addresses, and contact information when paying in an application that offers Apple Pay as a way of payment

How to set up the Touch ID sensor.

If you like to set up Touch ID, Go to Settings > Touch ID & Passcode. Set whether you need to use a fingerprint to unlock the iPhone and to make purchases. Tap Add a Fingerprint and then follow the on-screen commands. You can also add more than one fingerprint (for example, your thumb and forefinger, or one for your spouse).

Tips and tricks:
If you want to turn your iPhone off after setting up the Touch ID sensor, you'll be requested to confirm your passcode when you turn iPhone back on and unlock it for the first time. You'll similarly be asked for your original Apple ID password for the first purchase you made in the iTunes Store,

App Store, o ibook.
How to delete a fingerprint.
If you want to delete the fingerprint, tap it, and then remove Fingerprint. If you have registered more than one fingerprint, place your finger on the home button to find out which fingerprint it is.
To name a fingerprint.
If you want to name your fingerprint, tap the fingerprint, then enter a name, for instance, "Thumb." Use the Touch ID sensor of your iPhone to make a payment in the iTunes Store, App Store, or iBooks Store.

Whenever you are purchasing from the iTunes Store, App Store, or iBooks Store, follow the prompts instructions to enable your purchases with your fingerprint. Or go to your iPhone Settings > Touch ID & Passcode, and then turn on iTunes & App Store.

Using Touch ID for Apple Pay.
If you want to use the touch ID for Apple pay, Go to Settings > Touch ID & Passcode to make sure that Apple Pay is enabled with your Touch ID.

iCloud keychain
iCloud Keychain saves your website user names and passwords, credit card information, and Wi-Fi network information as well as up to date of your iPhone. iCloud Keychain function on all your approved devices (iOS 7 or later) Mac and PCs.

iCloud Keychain usually functions with Safari Password Generator and AutoFill. When setting up a new account,

Safari Password Generator recommends unique, difficult-to-guess passwords. You can also use AutoFill to have iPhone enter your user name and password info, making a login.

Tips and tricks:

Many websites do not support AutoFill. iCloud Keychain is a secured with 256-bit AES encryption during storage and transmission, and cannot be easily read by Apple.

To set up iCloud Keychain. Just go to Settings > iCloud > Keychain. Turn on iCloud Keychain, and then follow the on-screen guidelines. If you set up iCloud Keychain on another device, definitely, you must approve the use of your iCloud Keychain from one of the devices, or use your iCloud Security Code.

Tips: If you have ever forgotten your iCloud Security Code, you have to start over again and set up iCloud Keychain once more.

How to set up AutoFill.

If you like to set up autofill, go to Settings > Safari > Passwords & AutoFill. You should make sure that the names, passwords, and Credit Cards, are turned on (because they're always on the default setting). To add your credit card info, tap Saved Credit Cards.

The security code of your credit card is not saved automatically—you have to enter it manually here.

To automatically fill in your names, passwords, or credit card information on sites that accept it, tap a text field, and then tap AutoFill. To protect and safeguard your personal information, it is better to set up a passcode if you turn on iCloud Keychain and AutoFill.

How to Limit Ad Tracking

In order to restrict or reset Ad Tracking. Go to Settings > Privacy > Advertising. Turn on Limit Ad Tracking to stop apps from accessing your iPhone advertising identifier.

How to Find My iPhone

Find My iPhone can support you locate and secure your iPhone using the free Find My iPhone app, which is available in the App Store, on another iPhone, iPad, or iPod touch, or using a Mac or computer web browser signed in to www.icloud.com. Find My iPhone contains Activation Lock, which is designed to stop anyone else from using your iPhone if you ever lose it. Your Apple ID and password are essential to turn off Find My iPhone or to delete and reactivate your iPhone

To turn on Find My iPhone. Go to Settings > iCloud > Find My iPhone.

Tips and tricks:

To use Find My iPhone features, Find My iPhone must be turned on always before your iPhone is lost. Your iPhone must be able to connect to the Internet for you to discover and secure the device. Your iPhone sends its last location preceding to the battery running out when Send Last Location in Settings is turned on.

• Play Sound: Play a sound at maximum volume for two minutes, even if the ringer is set to silent.

• Lost Mode: Instantly lock your missing or theft iPhone with a passcode and send it a message displaying a contact number. iPhone tracks and reports its location, so you can also see where it's been when you check the Find My iPhone app. Lost Mode suspends the use of your credit and debit cards used for Apple Pay.

• *Erase iPhone*: to protect your privacy by erasing all the detail information and media on the iPhone and restoring it to its factory settings. Erase iPhone can also remove the ability of anyone to make payments using your credit and debit cards used for Apple Pay.

Tips and tricks;

Before giving away or selling your iPhone, you should erase it to remove all of your personal information and turn off Find My iPhone so that to ensure that the next user can activate and use the device normally. to do that just go to Settings > General > Reset > Erase All Content and Settings.

Charge the iPhone battery.

Connect iPhone11 to a power outlet using the included cable and USB electric power adapter.

You may charge the battery by connecting the iPhone to your PC, which also allows you to sync the iPhone with iTunes. Unless your keyboard has a high power USB 2.0 or 3.0 port, you should connect your iPhone to a USB 2.0 or 3.0 port on your PC.

Tips and tricks:

The iPhone battery sometime may drain instead of charge if the iPhone is connected to a computer that's turned off or is in standby or sleep mode.

To see the percentage or proportion of battery used by each app. Go to Settings > General > Usage, and then tap Battery Usage.

The battery icon always in the upper-right corner displays the battery charging status. To show the percentage of battery charge remaining, go to Settings > General > Usage. When syncing or using the iPhone, it can take longer to charge the battery.

Tips: If your iPhone is very low on power, it may show an image of a nearly depleted battery, indicating that your iPhone needs to charge for at least ten minutes before you can use it. If the iPhone battery is extremely low on power, the display may even be blank for up to two minutes before the low-battery image display.

Your iPhone rechargeable batteries have some degree of some charge cycles and may eventually need to be replaced when drained. You should take note that the iPhone battery isn't user-replaceable; it should be replaced only by an authorized Apple service provider.

Chapter Sixteen

Travel with your iPhone

If you wish you travel outside yours network area, you can avoid roaming charges by turning off voice and data services in Settings > Cellular. Some airlines allow you to keep your iPhone turned on if you switch to Airplane Mode. But remember you can't make calls or use Bluetooth. On the other hand, you can also listen to music, watch videos, play games, or use other applications that don't need network or phone connections. If the airline permits it, you can turn Wi-Fi or Bluetooth back on to enable those services, even while in Airplane Mode.

To turn on Airplane Mode. Swipe up from the bottom of the screen to open your iPhone Control Center and tap. You can turn Airplane Mode on or off in Settings. When Airplane Mode is on, it appears in the status bar at the top of the screen. You can turn Wi-Fi and Bluetooth either on or off in the Control Center of your iPhone.

Make requests with seri

For more suggestions, ask Siri "what want you do," or tap. Depending on the request, the on-screen response from Siri usually includes information, or Siri allows you to speak to your iPhone in order to send messages, schedule meetings, place phone calls. Siri comprehends natural speech, so you don't have to learn special commands or keywords. Just ask Siri anything, from "set the timer for four minutes" to "what movies are showing now?" Open applications, and turn features like Airplane Mode, Bluetooth, Do Not Disturb, and VoiceOver on or off. Siri is great for keeping you updated with the latest sports news, helping you decide on searching the iTunes Store or App Store for purchases.

Call for Siri.

If you want, call siri, press and hold the Home button until you heard Siri beeps, and then make your request. Control when Siri listens. Instead of letting Siri knows when you stop talking, you should continue to hold down the screen Home button while you speak and release it when you finish, Hey Siri. With your iPhone connected to a power source, you can use Siri without even pressing the Home button. Just say, "Hey Siri," then say your request. If you want to turn Hey Siri on or off, go to Settings > General > Siri > Allow "Hey Siri".

If you're using your iPhone headset, you can use the center or call button in place of the Home button.

How to response from Siri

Regularly you can tap the screen for more info or further action.

Tap to speak to Siri.

You can also tap for additional detail, or to perform some other action such as searching the web or opening a related app.

To change gender voice for Siri. Go to Settings > General > Siri (these may not be available in all regions).

Adjust the volume for Siri. Use the phone volume buttons while you're interacting with Siri.

Siri and apps work with several types of apps on the iPhone, including Phone, Messages, Maps, Clock, Calendar, and more. For instance, you can say things like:

- "Call Dad at home."
- "Do I have any new texts from Knoll?"

- *"I'm running low on gas."*
- *"Set an alarm for 7 a.m."*
- *"Cancel all my meetings on Saturday"* More examples of how you can use Siri with apps appear throughout this book.

How to tell Siri about yourself

If you want to tell Siri more about yourself—including things like your work and home addresses, and your spouse relationships—you can get personalized service like, "remind me to call my wife when I get home."

To tell Siri who you are.

 Fill out your detail information card in Contacts, then go to Settings > General > Siri > My Info and tap your name.

To let Siri know about your family relationship, say something like, "Diana John is my wife."

Make a corrections

If Siri doesn't get you right, you can tap to edit your request. Or tap again, and then clarify your request verbally. If you want to cancel the last command? Say "cancel," then tap the Siri icon, or press the screen Home button.

Siri Eyes Free

With Siri Eyes Free app, you can use iPhone features in your car without watching at or touching iPhone—you can control the seri entirely by speaking alone. To talk with Siri, just press and hold the voice command button on your steering wheel until you hear the Siri tone. You can ask Siri to call people, choice and play music, hear and compose text messages, get map directions, read your notifications, find calendar information, add reminders, and much more. Siri

Eyes Free is always available on select automobiles.

Use Siri Eyes Free.
You should connect your iPhone to your car using Bluetooth. Here it is better to refer to the user guide that came with your car.

Option for Siri settings

To set options for Siri, just go to Settings > General > Siri. Options include:
- *Turning Siri on or off*
- *Turning Allow "Hey Siri" on or off*
- *Language*
- *Voice gender (this may not be available in all regions)*
- *Voice feedback*
- *My Info card Stop accessing Siri when your iPhone is locked. Go to Settings > Apple Touch ID & Passcode (iPhone models with Touch ID) or Settings > Passcode (other models). You can as well disable Siri by turning on restrictions.*

Phone calls
If you want to make a call —just double-click the Home button. Dial manually. Tap Keypad, enter the number and then tap a Call.

- *Paste a number to a keypad: Tap the screen above the keyboard, then tap Paste.*
- *Enter a soft (three-second) pause: Touch the "*" key until a comma appears.*

For making a call on your, iPhone is as simple as selecting a number in your contacts list or tapping one of your favorites or recent calls.

Add favorites.

If you want to add any Favorites, you can make a call with a single tap. To add someone to your Favorites. You can add names to Favorites from your iPhone Contacts. In the Contacts, tap Add to Favorites at the bottom of a card and tap the number you want to add.

Delete a name from your Favorites list. Tap Edit. Return a recent call. Tap recent, and then tap the call. Tap to get more information about the call, or the caller. A red badge shows the number of missed calls. You can as well reach recent and favorite people you've been in contact with from the multitasking.

- Enter a hard pause (to pause dialing till you tap the Dial button): Touch the "#" key until a semicolon appears.
- Redial the last number: when you want to redial the last number, tap Keypad, tap call to display the number, and then tap call again. Use Siri or Voice Control. Just press and hold the Home button, say "call" or "dial," then say the number or name. You can also add "at home," "work," or "mobile.

You can say things like:
- "Call Diana's mobile."
- "Call the emergency department."
- "Redial that last number" When voice dialing a number, speak each digit separately—for example, "four one five, six two five...." For the 800 area code, if in the U.S., you can say

"eight hundred." Call over Wi-Fi. (this available in all regions).

iPhone Route calls over

Wi-Fi, go to Settings > Phone, then turn on Wi-Fi Calling. If the Wi-Fi connection is lost, calls switch automatically to your services' cellular network using VoLTE (Voice over LTE), if available. (VoLTE calls to switch to Wi-Fi when a Wi-Fi connection becomes readily available.) In previous models, a call is dropped if you lose the Wi-Fi connection. Contact your carrier for feature availability.

When somebody calls Tap Accept to answer an incoming call. Or if iPhone is locked, drag the slider. You can press the center button on your headset.

To Silence a call. Press the Sleep/Wake button or volume button. However, you can still answer the call after silencing it until it goes to voicemail.

If you want to decline a call and send it directly to voicemail. Do one of the following:

- Press the Sleep/Wake button twice.
- Press and hold the center button on your iPhone headset for about two seconds. Two beeps confirm that the call was declined.
- Tap Decline (if your iPhone is in awake Mode when the call comes in).

Tips and tricks:

In some regions, declined calls disconnected without being sent to voicemail.

Answer with a text message. Tap Message, and then select a reply or tap Custom. To create your replies, go to Settings >

Phone > Answer with Text, then tap the default messages and replace it with your text.

Remind yourself to return the call. Tap Remind Me and then indicate when you want reminded.

To make and receive calls on iPad, iPod touch, or Mac PC Continuity (iOS 8 or later) allows you make and receive calls on your iPad or iPod touch or on your Mac (with OS X Yosemite). Calls are relayed over your iPhone, which must be turned on and connected to an area's cellular network.

Other iOS device or Mac must always be connected to the Wi-Fi network and signed in to FaceTime and iCloud using the similar Apple ID as your iPhone.

Enable or disable iPhone calls through iPhone. On your iPhone, go to Settings > FaceTime and then turn iPhone Cellular Calls on or off.

Tips and tricks:

If available on your iPhone, Allow Wi-Fi Calls must be off. Just Go to Settings > Phone > then Wi-Fi Calls.

• Turn iPhone mobile Calls on or off for one more iOS device: On the device, go to Settings > FaceTime.

• Turn iPhone calls on or off for your Mac: On your Mac, open FaceTime, then select FaceTime > Preferences > Settings.

Receive a call on iPad, iPod touch, or Mac. To do that, just Swipe or click the notification to answer, ignore, or respond with a message.

Make a call from iPad, iPod touch, or Mac. Click a phone number in iPhone Contacts, Calendar, FaceTime, Messages, or Spotlight. You can tap a phone number from a recent contact in the multitasking screen on your iOS device.

Keep it quiet

If you wish to stay offline or a while? Just swipe up from the bottom edge of the screen to open Control Center and then turn on Do Not Disturb or Airplane Mode.

How to block unwanted callers?. On your contact card, tap Block this Caller (you can see a caller's contact card from your Favorites list or recent by tapping). You can as well block any callers in Settings > Phone > Blocked. You will not receive text message voice calls, FaceTime calls, or text messages from blocked callers.

When you're on a call, the screen displays some call options.

Mute your line. Or touch and hold to put a call on hold.

Mute your line. Or touch and hold to put your call on hold.

Make another call.

Dial or enter a number.

Use the speakerphone or a Bluetooth device.

Get the contact info.

End a call. Tap or press the Sleep/Wake button.

Use an app while on a call. Press the Home button, and then open the application. To return the call, tap the green bar at the top edge of the iPhone screen.

Respond to a second call. You can respond to a second call by:

• To ignore any call and send it to voicemail: Tap Ignore.

• To put the first call on hold and answer the new call: Tap holds + Accept.

•To end the first call and answer the new call: When using a

GSM network service, tap End + Accept. But With a CDMA network, tap End and when the second call rings back, tap Accept, or drag the slider when your iPhone is locked. With a call on hold, just tap Swap to switch in between calls or tap Merge Calls to talk with both at once. For more detail, see Conference calls below.

Tips and tricks:
With the CDMA network, you can't switch between calls if the second call was outgoing, but you can merge the calls. You can't merge calls if the second one was incoming. If you end the second call or the joined call, both calls are terminated.
Conference calls. Using GSM, you can set up a conference call with up to 5 people (this is depending on your service provider).
Tips: Conference calls may not be available if your call is using Voice over LTE (Vo LTE).
How to create a conference call. While you are on a call, tap Add Call, make another call, and then tap Merge Calls. Repeat to add more people to the conference.
  • Drop one person: Tap next to a person, and then tap End.
  • Talk privately with one person: Tap then taps Private next to the person. Tap Merge Calls to resume the conference.
  • Add an incoming caller: Tap Hold Call + Answer and then tap Merge Calls.

Chapter Seventeen

How to make emergency calls

*I*n order to make an emergency call when iPhone is locked. On the Enter Passcode iPhone screen, tap Emergency Call (to dial 911 if you are in the U.S.).
Tips: You can also use your iPhone to make an emergency call in any locations, provided that mobile network service is available, but you should not solely rely on it for emergencies. Some networks may not accept an emergency call from iPhone if the iPhone is not activated if iPhone is not compatible or configured to operate on a specific cellular network or (when applicable) if your iPhone does not have a SIM card or if the SIM card is PIN-locked.
In the United States, the location information is provided to emergency service providers when you dial 911. Please always review your service providers' emergency calling information to know the limits of emergency calling over Wi-Fi.
But with CDMA, when an emergency call ends, iPhone enters emergency call mode for some minutes to allow a call back from emergency services. During that time, data transmission and text messages are blocked.
To exit an emergency call mode (CDMA). Do one of the following:
• Tap the Back button on the screen.

- *Press the Sleep/Wake button on the Home button.*
- *Use the keypad to dial any non-emergency number.*

Visual voicemail

The visual voicemail allows you to see a list of your messages and select which one to listen to or delete, without having to walk through all of them. A badge on the Voicemail icon conveys to you how many unheard messages you receive. The first time you tap voicemail, you're prompted to create a new voicemail password and record your voicemail greeting.

Listen to a voicemail message. To listen to a voicemail, tap Voicemail, and then tap a message. To listen again, choose the message and tap. If visual voicemail isn't available with your network, tap Voicemail and follows the voice prompts.

Return the call on iPhone.

Speakerphone for Audio, when a Bluetooth device is connected. Just tap to select audio output.

The Messages in your iPhone are saved until you delete them, or your carrier services erase them.

Using Siri. Say something like:
- *"Do I have a new voicemail?"*
- *"Play the voicemail from Rose" Delete a message. Swipe or tap the message, and tap Delete.*

Tips: In some regions, deleted messages may be permanently erased by your service carrier. Your voice messages may be deleted if you change your SIM card.

To manage your deleted messages. Tap Deleted Messages at the end of the messages list, and then select:
- *Listen to a deleted message: just tap the message.*

- *Undelete a message: just tap the message and tap Undelete.*
- *Delete messages permanently: just tap to Clear All.*

Update your greeting. Tap Voicemail, tap Greeting, tap Custom, and then tap Record. Or, to use your carrier's basic greeting, tap Default. Set an alert sound for new voicemail. Go to Settings > Sounds. Change the password. Go to Settings > Phone > Change Voicemail Password.

Chapter Eighteen

Best iPhone 11 Screen Protectors in 2019

Getting a brand-new iPhone 11

Don't wait before you have the iPhone in hand to buy a good screen protector. Applying a new screen protector the moment you take away the protective film is the best way to avoid any fingerprints and dust. It is better to use one of these screen protectors to cover your brand-new iPhone from the first day. If you've started using your new iPhone and you like to add a screen protector, or you need to change a damaged one, just ensure to clean a screen thoroughly before applying. The iPhone XR and the iPhone 11 models have similar screen dimensions: 5.94-in-by-2.98-in. , an iPhone screen protector made for one will fit very well with others.

TETHYS Glass Screen Protector

Well-reviewed three-pack comes with a plastic frame to guide perfect installation in your iPhone. The black curved frame around the clear glass screen protector goes all the way to the edge of the screen and melts into the iPhone's bezel. Thus, you won't notice it once the protector is on screen. TETHYS deals with a lifespan warranty.

Trianium Screen Protector

What a good bargain! Just get three clear glass 0.25-millimeter screen protectors, besides for an installation frame to guarantee perfect placement. Trianium also offers a lifetime warranty on its screen protectors.

Irshe Screen Protector

This 9H hardness glass screen protector will avert scratches on the iPhone screen. There is a black frame around the edge that disappears into the iPhone's bezels; thus, you will not notice it. The frame makes lining up the phone screen protector for installation cooler.

LEADSTAR Privacy Screen Protector

If you don't want other people looking over your shoulder and onto your screen, you'll need to pick up this 2-pack of privacy screen protectors from LEADSTAR.

IRSHE Transparent Screen Protector

This screen protector offering from IRSHE does not have the black frame that supports with placement. Better, it's designed to leave room for the heavy-duty cases that have a wider lip.

Camera protection guide

FilmHoo Camera Lens Protector

To Protect those elegant camera lenses on the back of the new iPhone 11. FilmHoo Camera Lens Protector will not affect the camera's flash, resolution, and clarity.

All in one: Temdan Waterproof iPhone Case

If you prefer more than just a glass screen protector on the front of your phone, this tank is a waterproof case and iPhone screen protector in one.

TORRAS iPhone 11 Screen Protector

Get two in a package as one, but you get an additional one-year warranty if your screen protector cracks or gets scratched up.

Full-screen coverage: ESR iPhone 11 Screen Protector

Curved edges despicable your screen coverage is truly edge-to-edge. This two-pack contains a whole kit for perfect installation.

Spigen Tempered Glass Screen Protector

This two-pack of 9H-hardness case-friendly iPhone screen protectors come with a tray for cool installation.

Popular: Mkeke iPhone 11 Screen Protector

This common three-pack of case-friendly screen protectors come with a frame for easy installation.

Anti-glare:
OtterBox Increase Screen Protector with Glare Guard

OtterBox offers this premium iPhone screen protector with Glare Guard for improved screen clarity in all lighting situations, especially bright sunlight.

Totallee Tempered Glass Screen Protector
Isn't the low-priced screen protector on the list, but it's the one that kept my earlier iPhone's 11 screen pristine. This edge-to-edge screen cover was very easy to install, straight, and bubble-free. When installed, it is scarcely perceptible.

Caudabe Crystal-Shield Glass iPhone Screen Protector
The rounded edges of this high-end iPhone screen protector help it melt right into your iPhone once it's in place. Though the price isn't super low, you get two. The crystal-clear glass is fortified and smudge-resistant.

Clear plastic: Supershieldz for iPhone 11
If you don't like the weight and breakability of glass, this's another option. This six-pack of pure, high-definition PET film screen protectors offer lightweight protection for your new iPhone's screen.

Matte plastic: iCarez Matte Screen Protector
Some prefer matte plastic film, It's lightweight, smooth, and it reduces both fingerprints and screen glare. I think it cuts down the clarity of your screen. It's a useful trade-off. You can get six for one low price with iCarez.

Matte glass: Mothca Matte iPhone Screen Protector,
The silky-smooth matte screen experience in a solid, protective 9H tempered glass protector. It cuts glare and fingerprints are seldom noticeable. There's only one in the

package, but it has a lifespan warranty.

If you need any privacy screen, it is better to go for the LEADSTAR Privacy Screen Protector. This tinted can windows of the iPhone world. You may still see your phone screen, but the person was sitting closer to you cannot. Once more, you get two, so you have that spare if you need it.

Best iPhone Apps
Best applications for taking Night Mode photos on iPhone

These apps may help you get the Night mode effect, even on your older iPhones. With iPhone 11 Pro and Pixel 3 shooting in Night mode since Apple recently added Night mode to the iPhone 11, a lot of people have been praising this new camera feature. It has lastly made it possible to get wonderful low-light shots in complete darkness. What if you are using an old modern iPhone that is sans Night mode, and you don't plan on getting a new one right now? Don't feel any uneasiness, there are some handy apps out there that will help you capture great Night mode photos without having to buy a new iPhone for a few hundred dollars.

NeuralCam
This App came out very recently, but it's gotten a spot on this list because it's one of the best apps out there.
The function of NeuralCam is to blend computational photography and machine learning to bring the maximum of your low-light photos. The App captures a definite number of

frames as you hold the phone steady for some seconds. It uses advanced image processing algorithms and learning to elegantly merge the frames and brighten up the image colors, so you get a single high-quality image.

The whole process takes a few seconds, depending on the device that you're using. During these few seconds, you must hold the phone as steady as you can, so if a tripod is available, it is recommended if you want the complete best shots. The final photo is an image that is clearer, brighter, sharper, more defined, and colorful.

Neural Cam works mutually with the rear and front-facing cameras, so you can get great low-light selfies without resorting to using flash. The app works on all iPhones range from iPhone 6.

Night Camera HD

For those who need an app that can take good low-light night photos, while having some other features, Night Camera HD is a good option.

With Night Camera HD, you can get manual setting selections for light exposure up to a second, as well as separate residual light and ISO improvement. With these, you get more clear images with less interference and noise because of the extended exposure time.

Night Camera HD similarly saves in high-resolution PNG format, and there is luminosity. RGB histograms to support you see when an image would be over or underexposed. There's as well a self-timer, various aspect ratios, and a full-screen mode. For those who want to zoom, there is a 6x digital live zoom. Night Camera HD will get you a great night mode images.

NightCap Camera

NightCap is a new added impressive App that helps you get worthy night photos, with or without an iPhone 11.

With NightCap, you can be able not to take low-light and night photos, but you can record video and get 4K time-lapse videos. For astrophotographers, there is mate or mode, Stars Mode, Star Trails Mode, which is great for investigation. It allows for 4x higher ISO with the ISO Enhancement, so you get brighter low-light photos with less noise.

If you are defective in some astro-photography with iPhone, then NightCap is an excellent addition that can get you possibly photos of the stars, with your iPhone.

ProCam 7

ProCam is a great third-party camera replacement app overall, and it does have a genuinely dedicated night mode of the program that is comparable to what you get with your iPhone 11.

With ProCam's Night Mode, the shutter speed develops reduced to allow extra light to get captured by the sensor. You'll have four shutter speed types to select from on the menu. As you use Night Mode in ProCam, you'll need to hold the phone stable while it is taking the photo, which shouldn't be more than a few seconds. Else, you guessed it — the result will be shadowy.

Night Mode with ProCam is one of the powerful features that the App has. You can use in ProCam take in long exposure shooting mode, overexposure cautions, live histograms, 4K video recording, RAW, manual controls, focus peaking, and much more.

ProCam is absolutely a powerful camera app that several budding iPhone photographer should have in their list.

Cortex Camera app

As some of the apps, Cortex takes quite a few dozen exposures to create a high-resolution image that is free of noise and distortion. The exposures can be everywhere from 2-10 seconds long, depending on the low-light conditions, and need you to hold the phone steady. While a tripod is not required, it will certainly help and recommended if you want to get the best results.

Cortex Camera is capable of long exposure effects like motion blur and soft water. All photos taken with Cortex are saved with RAW data when available, so you get extreme sharpness and detail. You can also toggle shutter priority, ISO priority, and complete manual controls.

Conclusions

Thank you again for downloading this book!

I hope you have enjoyed this book and we hope that you are going to enjoy your iPhone maximally and efficiently. Finally, Please take the time to share your thoughts and post a review on Amazon. I would greatly appreciate!

Thank you and good luck!

About the Author

Philip Knoll *is CEO of techguideblog, the publishing company that published several IT books. He worked at Inte-route, Europe's largest voice and data network provider. Before Inte-route, he was working as a senior network engineer for Globtel Internet, a significant Internet and Telephony Services Provider to the market. He has been working with Linux for more than 10 years putting a strong accent on security for protecting vital data from hackers and ensuring good quality services for internet customers. Moving to VoIP services he had to focus even more on security as sensitive billing data is most often stored on servers with public IP addresses. He has been studying QoS implementations on Linux to build different types of services for IP customers and also to deliver good quality for them and for VoIP over the public Internet. Philip has also been programming educational software's with Perl, PHP, and Smarty for over 7 years mostly developing in-house management interfaces for IP and VoIP services.*

The link; bonus books

https://techguideblog.net/free-ebook-60-minutes-apple-watch-guide/

Our website is http//www.techguideblog.net

You should check it out and let me know what you think. I keep a blog there for our efficient interaction. I like to invite you follow my journey, by signing up my free newsletter. If you subscribed you get free copy of my books.mp3, pdf files, and tutorials

The list of my favorite online tools, plus notification of free future kindles book and offers. Pleases, if you're interested signup.

IPHONE 11 2020 EDITION

www.ingramcontent.com/pod-product-compliance
Lightning Source LLC
Chambersburg PA
CBHW021412210526
45463CB00001B/339